T0325587

Environmentally Conscious Materials Handling

Environmentally Conscious Materials Handling

Edited by
Myer Kutz

JOHN WILEY & SONS, INC.

This book is printed on acid-free paper.⊗

Published by John Wiley & Sons, Inc., Hoboken, New Jersey
Published simultaneously in Canada

Library of Congress Cataloging-in-Publication Data:

Environmentally conscious materials handling / edited by Myer Kutz.
p. cm.
Includes index.
ISBN 978-0-470-17070-0 (cloth)
1. Materials handling. 2 Green technology. I. Kutz, Myer.
TS180.E58 2009
658.7'810286—dc22

2009004204

Printed in the United States of America

10 9 8 7 6 5

To the Wistreich Family

Contents

Contributors

Shoou-Yuh Chang
North Carolina A&T State University
Greensboro, North Carolina

Banu Ekren
University of Louisville
Louisville, Kentucky

Mujde Erten-Unal
Old Dominion University
Norfolk, Virginia

Audeen Walters Fentiman
Purdue University
West Lafayette, Indiana

Kasper Hallenborg
The Maersk Mc-Kinney Moller Institute
University of Southern Denmark
Odense, Denmark

Sunderesh S. Heragu
University of Louisville
Louisville, Kentucky

Maria E. Mayorga
Clemson University
Clemson, South Carolina

Patrick Patterson
Texas Tech University
Lubbock, Texas

James L. Smith
Texas Tech University
Lubbock, Texas

Ravi Subramanian
Georgia Institute of Technology
Atlanta, Georgia

Berrin Tansel
Florida International University
Miami, Florida

Blake P. Tullis
Utah State University
Logan, Utah

Jeffrey C. Wolstad
Texas Tech University
Lubbock, Texas

Preface

Many readers will approach this series of books in Environmentally Conscious Engineering with some degree of familiarity with, or knowledge about, or even expertise in one or more of a range of environmental issues, such as climate change, pollution, and waste. Such capabilities may be useful for readers of this series, but they aren't strictly necessary, for the purpose of this series is not to help engineering practitioners and managers deal with the effects of man-induced environmental change. Nor is it to argue about whether such effects degrade the environment only marginally or to such an extent that civilization as we know it is in peril, or that any effects are nothing more than a scientific-establishment-and-media-driven hoax and can be safely ignored. (Authors of a plethora of books, even including fiction, and an endless list of articles in scientific and technical journals, have weighed in on these matters, of course.) On the other hand, this series of engineering books does take as a given that the overwhelming majority in the scientific community is correct, and that the future of civilization depends on minimizing environmental damage from industrial, as well as personal, activities. At the same time, the series does not advocate solutions that emphasize only curtailing or cutting back on these activities. Instead, its purpose is to exhort and enable engineering practitioners and managers to reduce environmental impacts, to engage, in other words, in Environmentally Conscious Engineering, a catalog of practical technologies and techniques that can improve or modify just about anything engineers do, whether they are involved in designing something, making something, obtaining or manufacturing materials and chemicals with which to make something, generating power, transporting people and freight, handling materials anywhere in the chain between manufacturing operations, warehousing, and distribution, or handling and transporting both municipal and dangerous wastes.

Increasingly, engineering practitioners and managers need to know how to respond to challenges of integrating environmentally conscious technologies, techniques, strategies, and objectives into their daily work, and, thereby, find opportunities to lower costs and increase profits while managing to limit environmental impacts. Engineering practitioners and managers also increasingly face challenges in complying with changing environmental laws. So companies seeking a competitive advantage and better bottom lines are employing environmentally responsible methods to meet the demands of their stakeholders, who now include not only owners and stockholders, but also customers, regulators, employees, and the larger, even worldwide community.

Engineering professionals need references that go far beyond traditional primers that cover only regulatory compliance. They need integrated approaches centered on innovative methods and trends in using environmentally friendly processes, as well as resources that provide a foundation for understanding and implementing principles of environmentally conscious engineering. To help engineering practitioners and managers meet these needs, I envisioned a flexibly connected series of edited books, each devoted to a broad topic under the umbrella of Environmentally Conscious Engineering.

The intended audience for the series is practicing engineers and upper-level students in a number of areas—mechanical, chemical, industrial, manufacturing, plant, power generation, transportation, and environmental—as well as engineering managers. This audience is broad and multidisciplinary. Practitioners work in a variety of organizations, including institutions of higher learning, design, manufacturing, power generation, transportation, warehousing, waste management, distribution, and consulting firms, as well as federal, state and local government agencies. So what made sense in my mind was a series of relatively short books, rather than a single, enormous book, even though the topics in some of the smaller volumes have linkages and some of the topics might be suitably contained in more than one freestanding volume. In this way, each volume is targeted at a particular segment of the broader audience. At the same time, a linked series is appropriate because every practitioner, researcher, and bureaucrat can't be an expert on every topic, especially in so broad and multidisciplinary a field, and may need to read an authoritative summary on a professional level of a subject that he or she is not intimately familiar with but may need to know about for a number of different reasons.

The Environmentally Conscious Engineering series is comprised of practical references for engineers who are seeking to answer a question, solve a problem, reduce a cost, or improve a system or facility. These books are not a research monographs. The purpose is to show readers what options are available in a particular situation and which option they might choose to solve problems at hand. I want these books to serve as a source of practical advice to readers. I would like them to be the first information resource a practicing engineer reaches for when faced with a new problem or opportunity—a place to turn to even before turning to other print sources, even any officially sanctioned ones, or to sites on the Internet. So the books have to be more than references or collections of background readings. In each chapter, readers should feel that they are in the hands of an experienced consultant who is providing sensible advice that can lead to beneficial action and results.

The five earlier volumes in the series have covered mechanical design, manufacturing, materials and chemicals processing, alternative energy production, and transportation. The sixth series volume, **Environmentally Conscious Materials**

Handling, has linkages to some of those earlier volumes, particularly manufacturing and transportation. The nine chapters in this volume can be divided into three parts. The first part, consisting of four chapters, deals with the handling of materials in manufacturing, warehousing, and supply chains, with regard not only to the physical entities themselves, but also to the workers who handle them. The second part, consisting of three chapters, covers the handling of waste, both municipal solid waste and hazardous waste, and landfill management. The third part, consisting of two chapters, deals with transportation issues, with respect first to radioactive materials and second to pipelines.

I asked the contributors, located not only in North America, but also in Europe, to provide short statements about the contents of their chapters and why the chapters are important. Here are their responses:

Sunderesh S. Heragu (University of Louisville in Louisville, Kentucky), who along with Banu Ekren contributed the chapter, **Materials Handling System Design**, writes, "This chapter discusses traditional and newer material handling systems that are used in manufacturing systems as well as warehouses. Although the main function of a material handling system is to transport parts—a non–value added activity—it is nevertheless important because material handling is a vital link between manufacturing (or service) processes without which a manufacturing system or warehouse cannot operate."

James L. Smith (Texas Tech University in Lubbock, Texas), who along with Jeffrey C. Woldstad and Patrick Patterson, contributed the chapter on **Ergonomics of Manual Materials Handling**, writes, "This chapter explores the relationships of human capabilities and limitations to materials handling. Whether the worker is required to manually handle materials or to use mechanical assists or automation, the human component of materials handling should not be ignored. This chapter explores the development of manual materials handling guidelines and provides recommendations for considering the human element in the design of materials handling systems."

Kasper Hallenborg (University of Southern Denmark in Odense, Denmark), who contributed the chapter on **Intelligent Control of Material Handling Systems**, writes, "Manufacturers are facing new challenges that require a more flexible production environment. Multiagent technologies propose an approach to increase flexibility, robustness, and adaptability in dynamic environments. The chapter introduces the technology and describes basic communication and organization principles of multiagent systems. The principles are exemplified by two real cases from industry."

Maria E. Mayorga (Clemson University in Clemson, South Carolina), who together with Ravi Subramanian, contributed the chapter on **Incorporating Environmental Concerns in Supply Chain Optimization**, writes, "Motivated by increasing regulatory and market-driven environmental pressures that impact supply chain decision-making, this chapter outlines how related trends require conventional supply chain optimization approaches to be revisited. Specifically, using real-world examples, we describe how various legislative, economic, and social factors can be characterized within supply chain optimization models. Accordingly, we provide recommendations of value to practitioners."

Shoou-Yuh Chang (North Carolina A&T State University in Greensboro, North Carolina), who contributed the chapter on **Municipal Solid Waste Management and Disposal**, writes, "Municipal solid waste management becomes a complicated problem for the United States and other countries with public health, environmental and economical concerns. This chapter addresses the management processes and disposal alternatives in dealing with this public-sector problem."

Mujde Erten-Unal (Old Dominion University in Norfolk, Virginia), who contributed the chapter on **Hazardous Waste Treatment**, writes, "Different manufacturing and industrial processes generate hazardous waste. In addition, manufactured products are consumed throughout the society and lead to generation of hazardous waste by commercial, agricultural, institutional, and homeowner activities. Therefore, it is important to understand different technologies for treating hazardous wastes generated from these activities. In this chapter, the hazardous waste treatment methods are grouped and described under physical-chemical, biological, thermal and land disposal categories."

Berrin Tansel (Florida International University in Miami, Florida), who contributed the chapter on **Sanitary Landfill Operations**, writes, "Landfills are critical for most waste management strategies, because they are the simplest, cheapest, and most cost-effective method of disposing of waste. This chapter presents engineering considerations which should be incorporated during planning, design, operation, closure, and postclosure of landfills so that impacts to the environment (e.g., leachate and gas releases to the environment) are minimized or mitigated. Waste generation patterns and characteristics can be influenced by public education, local policies, and ordinances. Procedures for site selection, design, operation and closure of sanitary landfills are presented."

Audeen Walters Fentiman (Purdue University in West Lafayette, Indiana), who contributed the chapter on **Transportation of Radioactive Materials**, writes, "This chapter focuses on transportation of highly radioactive materials such as used nuclear fuel from commercial nuclear power plants and high-level waste resulting from reprocessing used nuclear fuel. Coverage includes sources, amounts, and current locations of nuclear wastes that will eventually need to be transported, regulations governing transportation of these materials, and descriptions of the types of casks used to transport highly radioactive materials. These wastes have been generated at more than 100 nuclear power plants and about a dozen Department of Energy facilities around the country and will need to be transported to central locations for treatment or disposal."

Finally, Blake P. Tullis (Utah State Universityin Logan, Utah) who contributed the chapter on **Pipeline System Hydraulics**, writes, "Pipeline design includes selecting the most appropriate pipe size, pipe material, pumps, valves, joint and seal type, corrosion protection, and operational procedures to minimize the potential for transient pressures that can burst or collapse the pipe. A sound pipeline design is essential for protecting the surrounding environment form the pipeline contents and vice versa."

That ends the contributors' comments. I would like to express my heartfelt thanks to all of them for having taken the opportunity to work on this book. Their lives are terribly busy, and it is wonderful that they found the time to write thoughtful and complex chapters. I developed the book because I believed it could have a meaningful impact on the way many engineers approach their daily work, and I am gratified that the contributors thought enough of the idea that they were willing to participate in the project. Thanks also to my editor, Bob Argentieri, for his faith in the project from the outset. And a special note of thanks to my wife Arlene, whose constant support keeps me going.

Myer Kutz
Delmar, NY

CHAPTER 1

MATERIALS HANDLING SYSTEM DESIGN

Sunderesh S. Heragu and Banu Ekren
University of Louisville
Louisville, Kentucky

1 INTRODUCTION[1]

Material handling systems consist of discrete or continuous resources to move entities from one location to another. They are more common in manufacturing systems compared to service systems. Material movement occurs everywhere in a factory or warehouse—before, during, and after processing. Apple (1977) notes that material handling can account for up to 80 percent of production activity. Although material movement does not add value in the manufacturing process, half of the company's operation costs are material handling costs (Meyers 1993).

[1] Many of the sections in this chapter have been reproduced from Chapter 11 of Heragu (2008), with permission.

Therefore, keeping the material handling activity at a minimum is very important for companies.

Due to the increasing demand for a high variety of products and shorter response times in today's manufacturing industry, there is a need for highly flexible and efficient material handling systems. In the design of a material handling system, facility layout, product routings, and material flow control must be considered. In addition, various other factors must be considered in an integrated manner. The next section describes the ten principles of material handling as developed by the Material Handling Industry of America (MHIA). It presents a guideline for selecting equipment, designing a layout, standardizing, managing, and controlling the material movement as well as the handling system. Another section describes the common types of material handling systems. This chapter also discusses types of equipment, how to select material handling equipment, an operating model for material handling, and warehousing issues. It ends with a case study that implements some of these issues.

2 TEN PRINCIPLES OF MATERIAL HANDLING

If material handling is designed properly, it provides an important support to the production process. Following is a list of ten principles as developed by the MHIA, which can be used as a guide for designing material handling systems.

2.1 Planning

A *plan* is a prescribed course of action that is defined in advance of implementation. In its simplest form, a material handing plan defines the material (what) and the moves (when and where); together, they define the method (how and who). Five key aspects must be considered in developing a plan:

1. The plan should be developed in consultation between the planner(s) and all who will use and benefit from the equipment to be employed.
2. Success in planning large-scale material handling projects generally requires a team approach involving suppliers, consultants when appropriate, and end-user specialists from management, engineering, computer and information systems, finance, and operations.
3. The material handling plan should reflect the strategic objectives of the organization, as well as the more immediate needs.
4. The plan should document existing methods and problems, physical and economic constraints, and future requirements and goals.
5. The plan should promote concurrent engineering of product, process design, process layout, and material handling methods, as opposed to independent and sequential design practices.

2.2 Standardization

Material handling methods, equipment, controls, and software should be standardized within the limits of achieving overall performance objectives and without sacrificing needed flexibility, modularity, and throughput. Standardization means less variety and customization in the methods and equipment employed. There are three key aspects of achieving standardization:

1. The planner should select methods and equipment that can perform a variety of tasks under a variety of operating conditions and in anticipation of changing future requirements.
2. Standardization applies to sizes of containers and other load-forming components, as well as operating procedures and equipment.
3. Standardization, flexibility, and modularity must not be incompatible.

2.3 Work

The measure of work is material handling flow (volume, weight, or count per unit of time) multiplied by the distance moved. Material handling work should be minimized without sacrificing productivity or the level of service required of the operation. Five key points are important in optimizing the work:

1. Simplifying processes by reducing, combining, shortening, or eliminating unnecessary moves will reduce work.
2. Consider each pickup and set-down—that is, placing material in and out of storage—as distinct moves and components of the distance moved.
3. Process methods, operation sequences, and process/equipment layouts should be prepared that support the work minimization objective.
4. Where possible, gravity should be used to move materials or to assist in their movement while respecting consideration of safety and the potential for product damage (see Figure 1.1).
5. The shortest distance between two points is a straight line.

2.4 Ergonomics

Ergonomics is the science that seeks to adapt work or working conditions to suit the abilities of the worker. Human capabilities and limitations must be recognized and respected in the design of material handling tasks and equipment to ensure safe and effective operations. There are two key points in the ergonomic principles:

1. Equipment should be selected that eliminates repetitive and strenuous manual labor and that effectively interacts with human operators and users. The ergonomic principle embraces both physical and mental tasks.

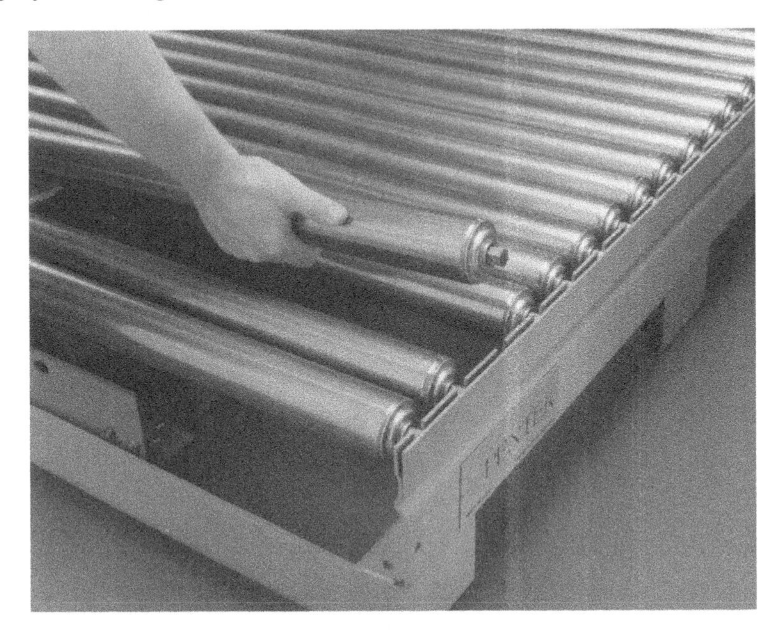

Figure 1.1 Gravity Roller Conveyor (Source: Courtesy of Pentek)

2. The material handling workplace and the equipment employed to assist in that work must be designed so they are safe for people.

2.5 Unit Load

A unit load is one that can be stored or moved as a single entity at one time, such as a pallet, container, or tote, regardless of the number of individual items that make up the load. Unit loads shall be appropriately sized and configured in a way that achieves the material flow and inventory objectives at each stage in the supply chain. When unit load is used in material flow, six key aspects deserve attention:

1. Less effort and work are required to collect and move many individual items as a single load than to move many items one at a time.
2. Load size and composition may change as material and products move through stages of manufacturing and the resulting distribution channels.
3. Large unit loads are common both pre- and postmanufacturing in the form of raw materials and finished goods.
4. During manufacturing, smaller unit loads, including as few as one item, yield less in-process inventory and shorter item throughput times.
5. Smaller unit loads are consistent with manufacturing strategies that embrace operating objectives such as flexibility, continuous flow, and just-in-time delivery.

6. Unit loads composed of a mix of different items are consistent with just-in-time and/or customized supply strategies as long as item selectivity is not compromised.

2.6 Space Utilization

Space in material handling is three-dimensional and therefore is counted as cubic space. Effective and efficient use must be made of all available space. This is a three-step process:

1. Eliminate cluttered and unorganized spaces and blocked aisles in work areas (see Figure 1.2).
2. In storage areas, balance the objective of maximizing storage density against accessibility and selectivity. If items are going to be in the warehouse for a long time, storage density is an important consideration. Avoid honeycombing loss (Figure 1.3). If items enter and leave the warehouse frequently, their accessibility and selectivity are important. If the storage density is too high to access or select the stored product, high storage density may not be beneficial.
3. Consider the use of overhead space when transporting loads within a facility. Cube per order index (COI) storage policy is often used in a warehouse. COI is a storage policy in which each item is allocated warehouse space based on the ratio of its storage space requirements (its cube) to the number of storage/retrieval transactions for that item. Items are listed in a nondecreasing order of their COI ratios. The first item in the list is allocated to the required number of storage spaces that are closest to the input/output (I/O) point; the second item is allocated to the required

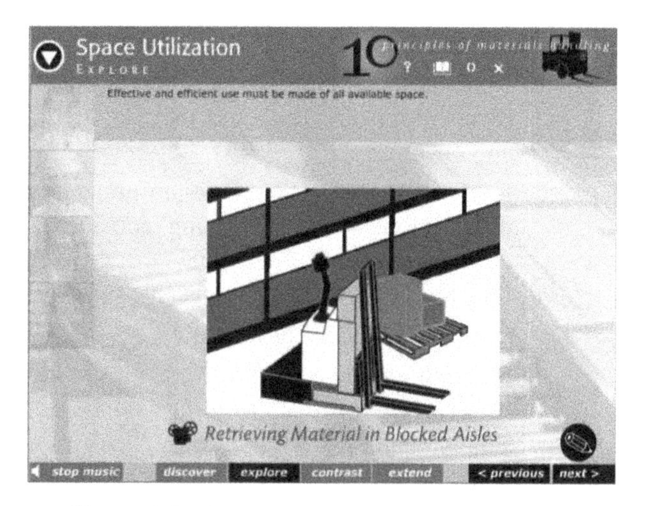

Figure 1.2 Retrieving material in blocked aisles

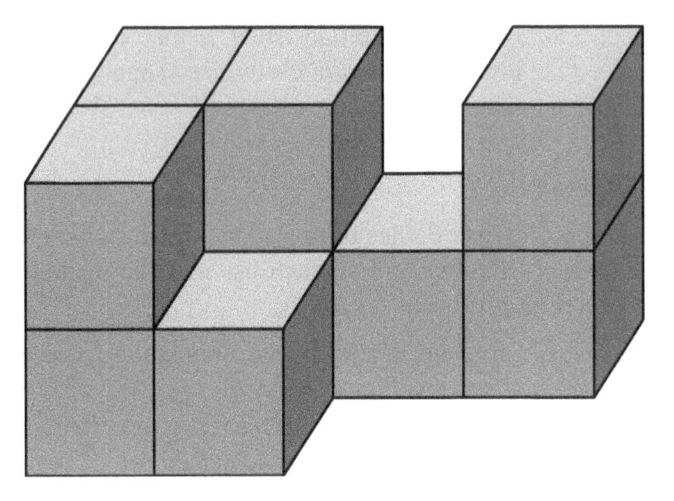

Figure 1.3 Honeycombing loss

number of storage spaces that are next closest to the I/O point, and so on. Figure 1.4 shows an interactive *playspace* in the "Ten principles of Materials Handling" CD that allows a learner to understand the fundamental concepts of the COI policy.

2.7 System

A *system* is a collection of interacting or interdependent entities that form a unified whole. Material movement and storage activities should be fully integrated to form a coordinated operational system that spans receiving, inspection, storage, production, assembly, packaging, unitizing, order selection, shipping, transportation, and the handling of returns. Here are five key aspects of the system principle:

1. Systems integration should encompass the entire supply chain, including reverse logistics. It should include suppliers, manufacturers, distributors, and customers.
2. Inventory levels should be minimized at all stages of production and distribution, while respecting considerations of process variability and customer service.
3. Information flow and physical material flow should be integrated and treated as concurrent activities.
4. Methods should be provided for easily identifying materials and products, for determining their location and status within facilities and within the supply chain, and for controlling their movement. For instance, bar coding is the traditional method used for product identification. Radio frequency identification (RFID) uses radio waves to automatically identify objects as

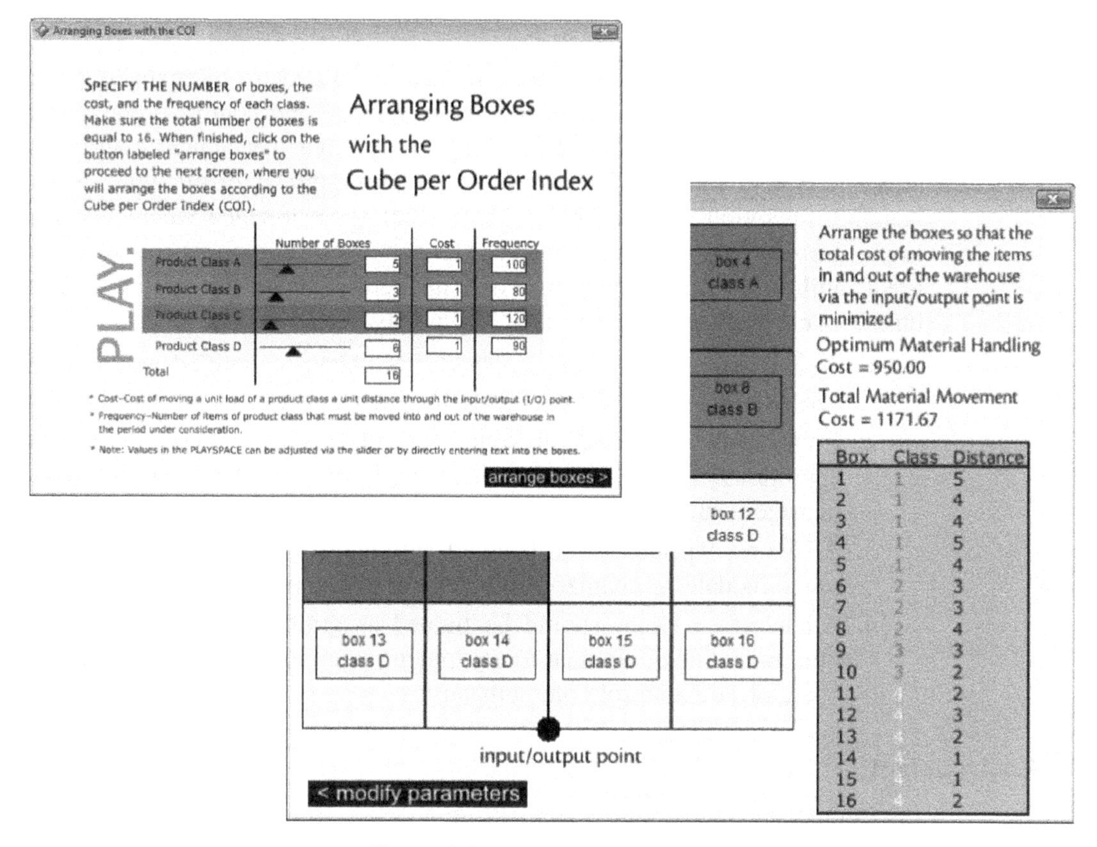

Figure 1.4 Example of COI policy

they move through the supply chain. The big difference between the two automatic data capture technologies is that bar coding is a line-of-sight technology. In other words, a scanner has to "see" the bar code to read it, which means people usually have to orient the bar code toward a scanner for it to be read. RFID tags can be read as long as they are within the range of a reader, even if there is no line of sight. Bar codes have other shortcomings, as well. If a label is ripped, soiled, or falls off, there is no way to scan the item. Also, standard bar codes identify only the manufacturer and product, not the unique item. The bar code on one gallon of 2 percent milk is the same as on every other gallon of the same brand, making it impossible to identify which one might pass its expiration date first. RFID can identify items individually.

5. Customer requirements and expectations regarding quantity, quality, and on-time delivery should be met without exception.

2.8 Automation

Automation is concerned with the application of electro-mechanical devices, electronics, and computer-based systems to operate and control production and service activities. It suggests the linking of multiple mechanical operations to create a system that can be controlled by programmed instructions. Material handling operations should be mechanized and/or automated where feasible to improve operational efficiency, increase responsiveness, improve consistency and predictability, decrease operating costs and eliminate repetitive or potentially unsafe manual labor. There are four key points in automation:

1. Preexisting processes and methods should be simplified and/or reengineered before any efforts at installing mechanized or automated systems.
2. Computerized material handling systems should be considered where appropriate for effective integration of material flow and information management.
3. All items expected to be handled automatically must have features that accommodate mechanized and automated handling.
4. All interface issues should be treated as critical to successful automation, including equipment to equipment, equipment to load, equipment to operator, and control communications.

2.9 Environment

Environmental consciousness stems from a desire not to waste natural resources and to predict and eliminate the possible negative effects of our daily actions on the environment. Environmental impact and energy consumption should be considered as criteria when designing or selecting alternative equipment and material handling systems. Here are the three key points:

1. Containers, pallets, and other products used to form and protect unit loads should be designed for reusability when possible and/or biodegradability as appropriate.
2. Systems design should accommodate the handling of spent dunnage, empty containers, and other byproducts of material handling.
3. Materials specified as hazardous have special needs with regard to spill protection, combustibility, and other risks.

2.10 Life Cycle

Life-cycle costs include all cash flows that will occur between the time the first dollar is spent to plan or procure a new piece of equipment, or to put in place a new method, until that method and/or equipment is totally replaced. A thorough economic analysis should account for the entire life cycle of all material handling equipment and resulting systems. There are four key aspects:

1. Life-cycle costs include capital investment, installation, setup and equipment programming, training, system testing and acceptance, operating (labor, utilities, etc.), maintenance and repair, reuse value, and ultimate disposal.
2. A plan for preventive and predictive maintenance should be prepared for the equipment, and the estimated cost of maintenance and spare parts should be included in the economic analysis.
3. A long-range plan for replacement of the equipment when it becomes obsolete should be prepared.
4. Although measurable cost is a primary factor, it is certainly not the only factor in selecting among alternatives. Other factors of a strategic nature to the organization that form the basis for competition in the marketplace should be considered and quantified whenever possible.

These ten principles are vital to material handling system design and operation. Most are qualitative in nature and require the industrial engineer to employ these principles when designing, analyzing, and operating material handling systems.

3 TYPES OF MATERIAL HANDLING EQUIPMENT

In this section, we list various equipments that actually transfer materials between the multiple stages of processing. There are a number of different types of material handling devices (MHDs), most of which move materials via material handling paths on the shop floor. However, there are some MHDs—such as cranes, hoists, and overhead conveyors—that utilize the space above the machines. The choice of a specific MHD depends on a number of factors, including cost, weight, size, and volume of the loads; space availability; and types of workstations. So, in some cases the MHS interacts with the other subsystems. If we isolate MHS from other subsystems, we might get an optimal solution relative to the MHDs but one that is suboptimal for the entire system.

There are seven basic types of MHDs (Heragu 2008): conveyors, palletizers, trucks, robots, automated guided vehicles, hoists cranes and jibs, and warehouse material handling devices. In this section, we will introduce the seven basic types of MHDs. In the following section, we will discuss how to choose the "right" equipment and how to operate equipment in the "right" way.

3.1 Conveyors

Conveyors are fixed-path MHDs. In other words, conveyors should be considered only when the volume of parts or material to be transported is large and when the transported material is relatively uniform in size and shape. Depending on the application, there are many types of conveyors—accumulation conveyor, belt conveyor, bucket conveyor, can conveyor, chain conveyor, chute conveyor, gravity conveyor, power and free conveyor, pneumatic or vacuum conveyor, roller conveyor, screw conveyor, slat conveyor, tow line conveyor, trolley conveyor, and wheel conveyor. Some are pictured in Figure 1.5. Our list is not meant to be complete, and other variations are possible. For example, belt conveyors may be classified as troughed belt conveyors (used for transporting bulky material such as coal) and magnetic belt conveyors (used for moving ferrous material against gravitational force). For the latest product information on conveyors and other types of material handling equipment, we strongly encourage the reader to refer to recent issues of *Material Handling Engineering* and *Modern*

(a)

Figure 1.5a Conveyors used in sortation applications (Source: Courtesy of Vanderlande Industries)

(b)

Figure 1.5b Accumulation conveyor (Source: Courtesy of Nike, Belgium)

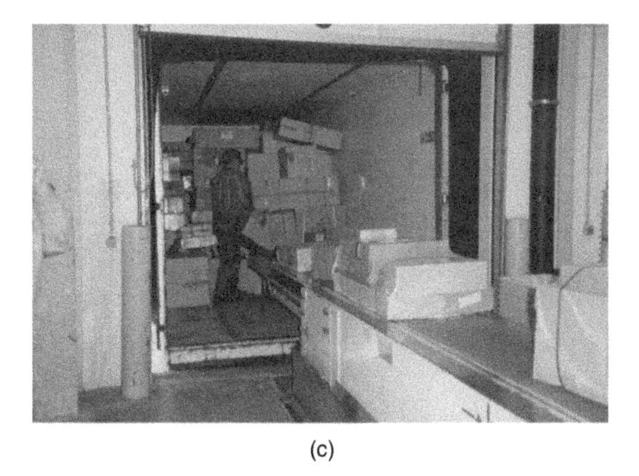

(c)

Figure 1.5c Extendable dock conveyor (Source: Courtesy of DPD, Germany)

(d)

Figure 1.5d Belt conveyor (Source: Courtesy of FKI Logistex)

Materials Handling. These publications not only have articles illustrating use of the material handling equipment but also numerous product advertisements.

3.2 Palletizers

Palletizers are high-speed automated equipment used to palletize containers coming off production or assembly lines. With operator-friendly touch-screen controls, they palletize at the rate of a hundred cases per minute (see Figure 1.6), palletize two lines of cases simultaneously, or simultaneously handle multiple products.

(e)

Figure 1.5e Chute and tilt-tray conveyor (Source: Courtesy of Dematic Corp.)

(f)

Figure 1.5f Overhead conveyor used in automobile assembly plant (Source: Courtesy of Gould Communications)

3.3 Trucks

Trucks are particularly useful when the material moved varies frequently in size, shape, and weight, when the volume of the parts or material moved is low, and when the number of trips required for each part is relatively small. There are several trucks in the market with different weight, cost, functionality, and other features. Hand truck, fork lift truck, pallet truck, platform truck, counter-balanced truck, tractor-trailer truck, and automated guided vehicles (AGVs) are some examples of trucks (see Figure 1.7).

Figure 1.6 High-speed palletizer (Source: Courtesy of FKI Logistex)

3.4 Robots

Robots are programmable devices that resemble the human arm. They are also capable of moving like the human arm and can perform functions such as weld, pick and place, load and unload (see Figure 1.8). Some advantages of using a robot are that they can perform complex repetitive tasks automatically and they can work in hazardous and uncomfortable environments that a human operator cannot work. The disadvantage is that robots are relatively expensive.

3.5 Automated Guided Vehicles

AGVs have become very popular, especially in the past decade, and will continue to be the dominant type of MHD in the years to come. The first system was installed in 1953, and the technology continues to expand. AGVs can be regarded as a type of specially designed robots. Their paths can be controlled in a number of different ways. They can be fully automated or semiautomated. AGVs are becoming more flexible with a wider range of applications using more diverse vehicle types, load transfer techniques, guide path arrangements, controls, and control interfaces. They can also be embedded into other MHDs. A sample of AGVs and their applications are illustrated in Figure 1.9.

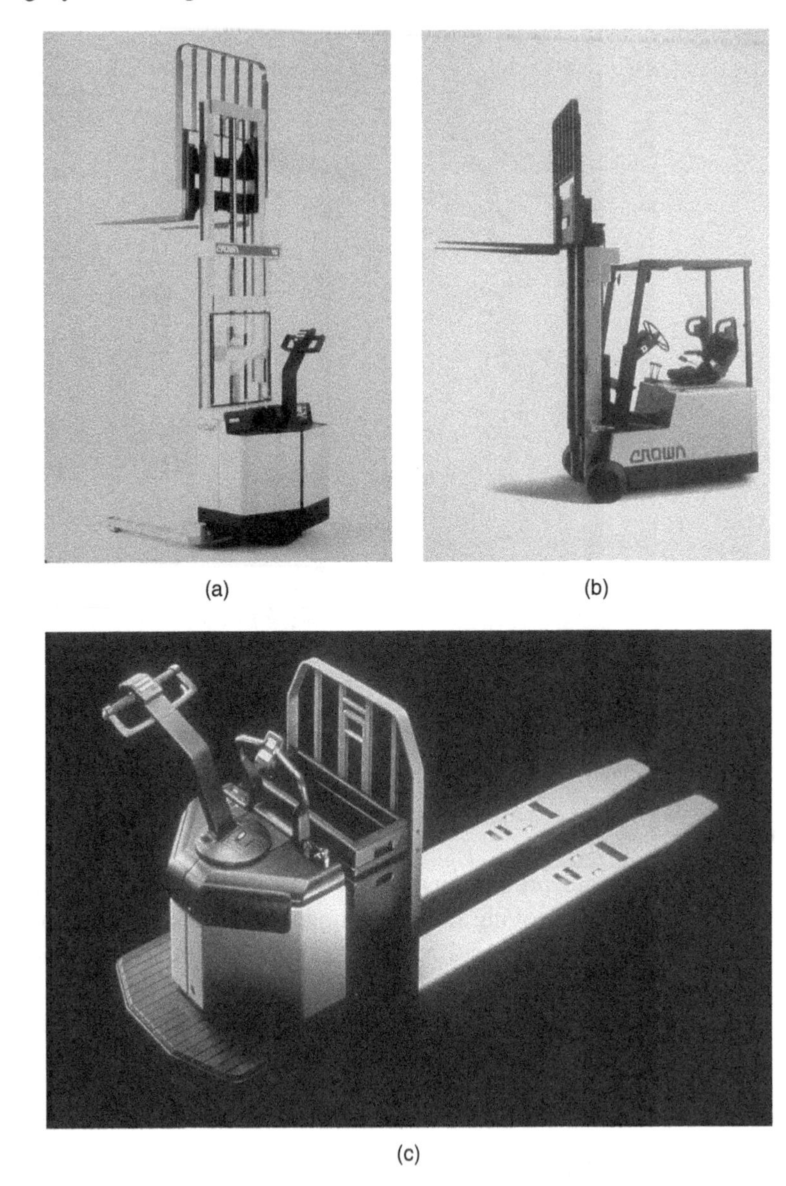

Figure 1.7a,b,c Examples of industrial trucks (Source: Courtesy of Crown Corporation)

3.6 Hoists, Cranes, and Jibs

These MHDs are preferred when the parts to be moved are bulky and require more space for transportation. Because the space above the machines is typically utilized only for carrying power and coolant lines, there is abundant room to transport bulky material. The movement of material in the overhead space does not affect production process and worker in a factory. The disadvantages

Figure 1.8 Use of robots in pick and place and welding operations (Source: Courtesy of Vanderlande Industries, Gould Communications, and Fraunhofer Institute, IML-Dortmund)

of these MHDs are that they are expensive and time-consuming to install (see Figures 1.10, 1.11, and 1.12).

3.7 Warehouse Material Handling Devices

These are typically referred to as storage and retrieval systems. If they are automated to a high degree, they are referred to as automated storage and retrieval systems (AS/RS). The primary functions of warehouse material handling

Figure 1.9 Use of AGVs in distribution and manufacturing activities (Source: Courtesy of Gould Communications)

devices are to store and retrieve materials as well as transport them between the pick/deposit (P/D) stations and the storage locations of the materials. An AS/RS is shown in Figure 1.13.

AS/RSs are capital-intensive systems. However, they offer a number of advantages, such as low labor and energy costs, high land or space utilization, high reliability and accuracy, and high throughput rates.

3.8 Autonomous Vehicle Storage and Retrieval System

Autonomous vehicle storage and retrieval systems (AVS/RS) represent a relatively new technology for automated unit load storage systems. In this system, the autonomous vehicles function as storage/retrieval (S/R) devices. Within the storage rack, the key distinction of AVS/R systems relative to traditional crane-based automated storage and retrieval systems (AS/RS) is the movement patterns of the S/R device. In AS/RS, aisle-captive storage cranes can move in the horizontal and vertical dimensions, simultaneously to store or retrieve unit loads. In an AVS/RS, vehicles use a fixed number of lifts for vertical movement and follow rectilinear

Figure 1.10 Manual, electric, and pneumatic hoists (Source: Courtesy of Harrington and Ingersoll-Rand)

Figure 1.11 Gantry cranes (Source: Courtesy of B.E. Wallace Products Corp. and Mannesmann Dematic)

Figure 1.12 AGV and gantry crane used for loading containers on ships (Source: Courtesy of Europe Combined Terminals B.V., The Netherlands)

flow patterns for horizontal travel. Although the travel patterns in an AS/RS are generally more efficient within storage racks (see Figure 1.13), an AVS/RS has a significant potential advantage in the adaptability of system throughput capacity to transactions demand by changing the number of vehicles operating in a fixed storage configuration (see Figure 1.14). For example, decreasing the number of

Figure 1.13 AS/RS (Source: Courtesy of Vanderlande Industries)

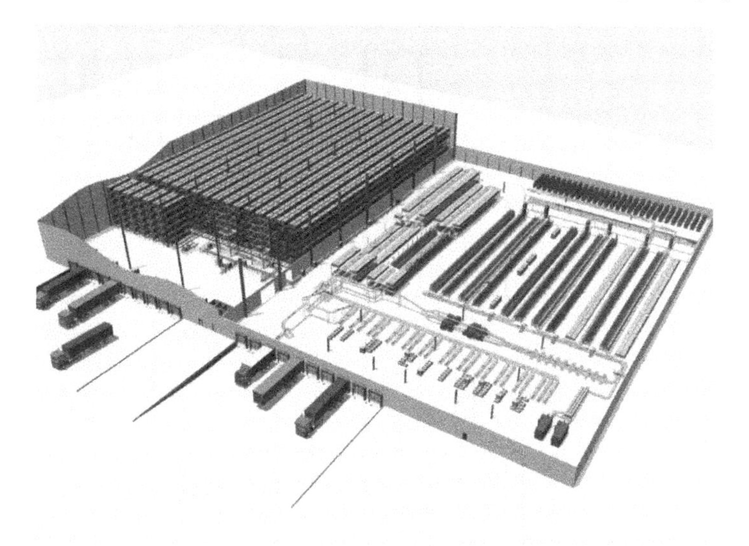

Figure 1.14 A typical AVS/RS (Source: Courtesy of Savoye Logistics)

vehicles increases the transaction cycle times and utilization, which are also key measures of system performance.

4 HOW TO CHOOSE THE "RIGHT" EQUIPMENT

Apple (1977) has suggested the use of the "material handling equation" in arriving at a material handling solution. The methodology illustrated in Figure 1.15 uses six major questions: why (select material handling equipment), what (is the material to be moved), where and when (is the move to be made), how (will the move be made), and who (will make the move). All these six questions are extremely important and should be answered satisfactorily.

The material handling equation can be specified as: *Material+Move=Method*, as shown in Figure 1.15. Very often, when the *material* and *move* aspects are analyzed thoroughly, it automatically uncovers the appropriate material handling *method*. For example, analysis of the type and characteristics of *material* may reveal that the material is a large unit load on wooden pallets. Further analysis of the logistics, characteristics and type of *move* may indicate that 6 meters load/unload lift is required, distance traveled is 50 meters, and some maneuvering is required while transporting the unit load. This suggests that a fork lift truck would be a suitable material handling device. Even further analysis of the method may tell us more about the specific features of the fork lift truck. For example, narrow aisle fork lift truck, with a floor load capacity of $^1/_2$ ton, and so on.

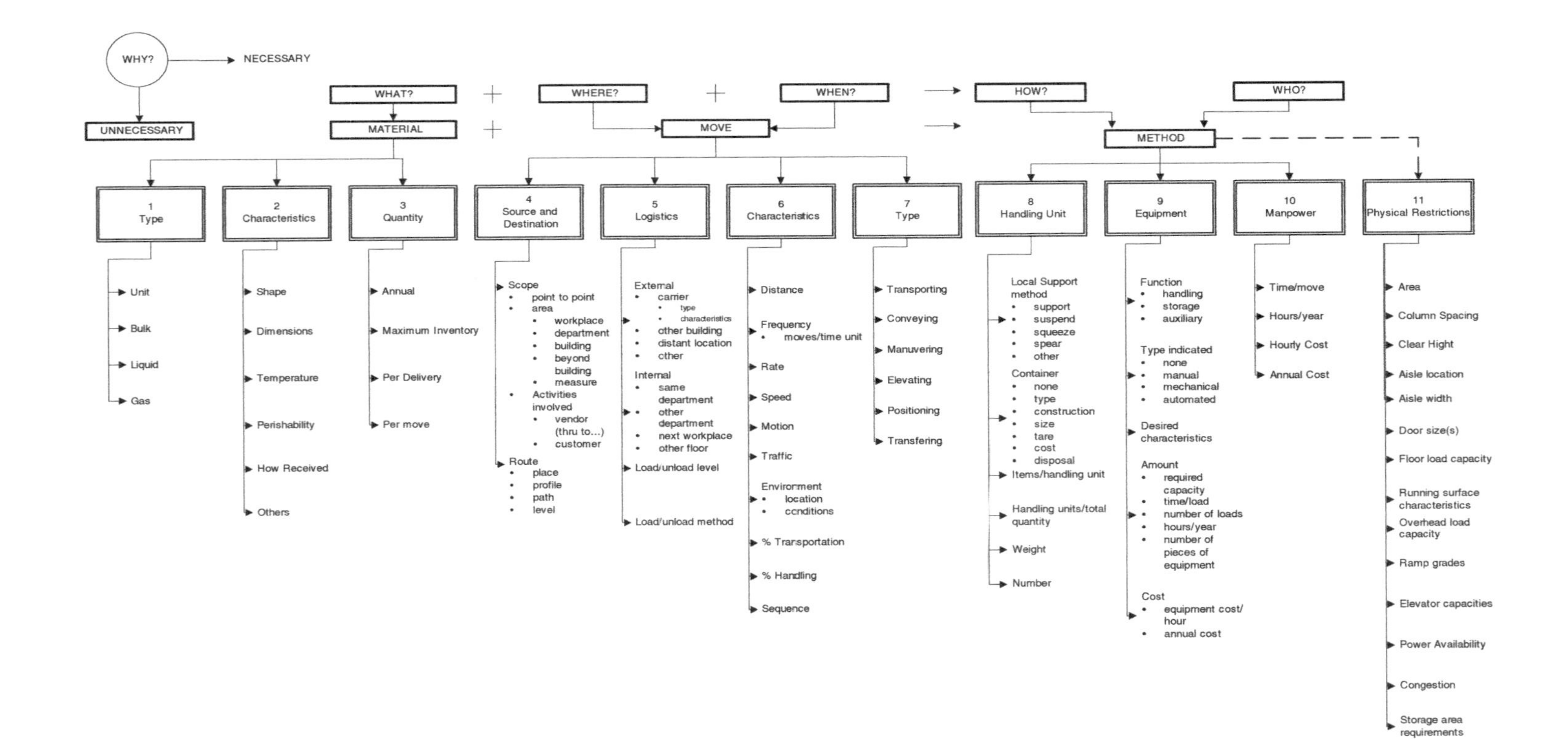

Figure 1.15 Material handling equation (Source: Courtesy of James M. Apple, Jr.)

5 A MULTIOBJECTIVE MODEL FOR OPERATION ALLOCATION AND MATERIAL HANDLING EQUIPMENT SELECTION IN FMS DESIGN

From both a conceptual as well as a computational point, only a few mathematical programming models have been proposed for the material handling system selection problem. Most of the studies have focused on material handling equipment optimization, rather than the entire material handling system. Sujono and Lashkari (2006) proposed a multiobjective model for selecting MHDs and allocating material handling transactions to them in flexible manufacturing system (FMS) design. They propose a model that integrates operation allocation (OA) and MHD selection problem. Their study is an extension of the Paulo et al. (2002) and Lashkari et al. (2004) studies. The main differences from the previous models are the new definition of the variables and the introduction of a new variable that links the selection of a machine to perform manufacturing operation with the material handling requirements of that operation. In addition, they include all the costs associated with material handling operations and suboperations, and the complete restructuring of the constraints that control the selection of the material handling equipment and their loading, in the objective function. Their model is presented as follows.

$h \in \{1, 2, \ldots, H\}$: major MH operations
$\hat{h} \in \{1, 2, \ldots, \hat{H}\}$: MH suboperations
$e \in E_{jh\hat{h}}\{1, 2, \ldots, E\}$: set of MH equipment that can handle the combination of MH operation/suboperation at machine j
$j \in J_{ips}\{1, 2, \ldots, m\}$: set of machines that can perform operation s of part type i under process plan p

Parameters

b_j : time available on machine j
OC_{ipj} : cost of performing operation s of part type i under process plan p on machine j (\$)
d_i : demand for part type i (units)
SC_j : setup cost of machine j (\$)
t_{ijp} : time for performing operation s of part type i under process plan p on machine j
$T_{ijh\hat{h}e}$: MH cost of performing the combination of MH operation/suboperation for part type i on machine j using MH equipment e (\$)
L_e : time available on MH equipment e
$I_{h\hat{h}e}$: time for MH equipment e to perform the combination of MH operation/suboperation
$\hat{W}_{it}$: relative weight of the product variable t on part type i
W_{et} : relative weight of the product variable t on MH equipment e

$W_{h\hat{h}e}$	: relative degree of capability of MH equipment e to perform the combination of MH operation/suboperation
C_{ei}	: compatibility between MH equipment e and part type i

Decision Variables

$Z(ip) \in \{1, 0\}$	: 1 if part type i uses process plan p; 0 otherwise
$Y_{sj}(ip) \in \{1, 0\}$	: 1 if machine j performs operation s of part type i under process plan p; 0 otherwise
$A_{ijph\hat{h}} \in \{1, 0\}$	: 1, if part type i under process plan p requires the combination of MH operation/suboperation at machine j; 0 otherwise
$X_{ijph\hat{h}e} \in \{1, 0\}$	: 1 if the combination of MH operation/suboperation requires MH equipment e at machine j where operation s of part type i under process plan p is performed; 0 otherwise
$M_j \in \{1, 0\}$	: 1 if machine j is selected; 0 otherwise
$D_e \in \{1, 0\}$	: 1 if MH equipment e is selected; 0 otherwise

The first part of the objective function is presented as equation (1):

$$F_1 = \sum_{i=1}^{n} d_i \sum_{p=1}^{P(i)} \sum_{s=1}^{S(ip)} \sum_{j \in J_{ips}} OC_{ipj} Y_{sj}(ip) + \sum_{j=1}^{m} SC_j M_j$$
$$+ \sum_{i=1}^{n} d_i \sum_{p=1}^{P(i)} \sum_{s=1}^{S(ip)} \sum_{j \in J_{ips}} \sum_{h=1}^{H} \sum_{\hat{h}=1}^{\hat{H}} \sum_{e \in E_{jh\hat{h}}} T_{ijh\hat{h}e} X_{ijph\hat{h}e}. \quad (1)$$

The second part of the objective function is formulated as equation (2):

$$F_2 = \sum_{e=1}^{E} \sum_{h=1}^{H} \sum_{\hat{h}=1}^{\hat{H}} W_{h\hat{h}e} \sum_{i=1}^{n} C_{ei} \sum_{p=1}^{P(i)} \sum_{s=1}^{S(ip)} \sum_{j \in J_{ips}} X_{ijph\hat{h}e}, \quad (2)$$

where

$$C_{ei} = 1 - \frac{\sum_{t=1}^{T} \left| W_{et} - \hat{W}_{it} \right|}{4T}.$$

Here, $T = 5$ and refers to the five major variables used to identify the dimensions of the characteristics mentioned by Ayres (1988). Integer numbers are used to assign values to the subjective factors, W parameters, W_{et}, $W_{h\hat{h}e}$ and $\hat{W}_{it}$. The rating scales range from 0 to 5 for W_{et} and $W_{h\hat{h}e}$ and 1 to 5 for $\hat{W}_{it}$ (Ayres 1988). A 5 for W_{et} means that the piece of equipment is best suited to handle parts with a very high rating of product variable t. A 0 means, do not allow this piece of equipment to handle parts with product variable t. A 5 for $W_{h\hat{h}e}$ means that it is excellent in performing the operation/suboperation combination. And a 0 means that it is incapable of performing the operation/suboperation combination. A 5 for $\hat{W}_{it}$ means that the part type exhibits a very high level of the key product variable t. And a 0 means that the part type exhibits a very low level of the key product variable t.

The first part of objective function's three terms indicates the manufacturing operation costs, the machine setup costs, and the MH operation costs, respectively. The second part of the objective function computes the overall compatibility of the MH equipment. As a result, the formulation of the problem is a multiobjective model seeking to strike a balance between the two objectives.

There are nine constraints in this model:

1. Each part type can use only one process plan:

$$\sum_{p=1}^{P(i)} Z(ip) = 1 \quad \forall i. \tag{3}$$

2. For a given part type i under process plan p, each operation of the selected process plan is assigned to only one of the available machines:

$$\sum_{j \in J_{ips}} Y_{sj}(ip) = Z(ip) \quad \forall i, p, s. \tag{4}$$

3. Once a machine is selected for operation s of part type i under process plan p, then all the $(h\hat{h})$ combinations corresponding to (sj) must be performed:

$$Y_{sj}(ip) = A_{sjh\hat{h}}(ip) \quad \forall i, p, s, j, h, \hat{h}. \tag{5}$$

4. Each $h\hat{h}$ combination can be assigned to only piece of available and capable MH equipment:

$$\sum_{e \in E_{jh\hat{h}}} X_{ijph\hat{h}e} = A_{ijph\hat{h}} \quad \forall i, p, s, j, h, \hat{h}. \tag{6}$$

5. At least one operation must be allocated to a selected machine:

$$\sum_{i=1}^{n} \sum_{p=1}^{P(i)} \sum_{s=1}^{S(ip)} \sum_{h=1}^{H} \sum_{\hat{h}=1}^{\hat{H}} \sum_{e \in E_{jh\hat{h}}} X_{ijph\hat{h}e} \geq M_j \quad \forall j. \tag{7}$$

6. The allocated operations cannot exceed the corresponding machine's capacity:

$$\sum_{i=1}^{n} d_i \sum_{p=1}^{P(i)} \sum_{s=1}^{S(ip)} \sum_{h=1}^{H} \sum_{\hat{h}=1}^{\hat{H}} \sum_{e \in E_{jh\hat{h}}} t_{sj}(ip) X_{ijph\hat{h}e} \leq b_j M_j \quad \forall j. \tag{8}$$

7. A specific MH equipment can be selected only if the corresponding *type* of equipment is selected:

$$D_e \leq D_{\hat{e}} \quad \forall e, \hat{e}. \tag{9}$$

8. Each MH equipment selected must perform at least one operation:

$$\sum_{i=1}^{n}\sum_{h=1}^{H}\sum_{\hat{h}=1}^{\hat{H}}\sum_{p=1}^{P(i)}\sum_{s=1}^{S(ip)}\sum_{j\in J_{ips}} X_{ijph\hat{h}e} \geq D_e \quad \forall e. \tag{10}$$

9. The MH equipment capacity cannot be exceeded:

$$\sum_{i=1}^{n} d_i \sum_{h=1}^{H}\sum_{\hat{h}=1}^{\hat{H}}\sum_{p=1}^{P(i)}\sum_{s=1}^{S(ip)}\sum_{j\in J_{ips}} X_{ijph\hat{h}e} \geq D_e \quad \forall e. \tag{11}$$

6 WAREHOUSING

Many manufacturing and distribution companies maintain large warehouses to store in-process inventories or components received from an external supplier. They are involved in various stages of the sourcing, production, and distribution of goods, from raw materials through the finished goods. The true value of warehousing lies in having the right product in the right place at the right time. Thus, warehousing provides the time-and-place utility necessary for a company and is often one of the most costly elements. Therefore, its successful management is critical.

6.1 Just-in-Time (JIT) Manufacturing

It has been argued that warehousing is a time-consuming and non–value-adding activity. Because additional paperwork and time are required to store items in storage spaces and retrieve them later when needed, the JIT manufacturing philosophy suggests that one should do away with any kind of temporary storage and maintain a pull strategy in which items are produced only as and when they are required. That is, they should be produced at a certain stage of manufacturing, only if they are required at the next stage.

JIT philosophy requires that the same approach be taken toward components received from suppliers. The supplier is considered as another (previous) stage in manufacturing. However, in practice, because the demand is continuous, that means that goods need to be always pulled through the supply chain to respond to demand quickly. The handling of returned goods is becoming increasingly important (e.g., Internet shopping may increase the handling of returned goods), and due to the uncertainty inherent in the supply chain, it is not possible to completely do away with temporary storage.

6.2 Warehouse Functions

Every warehouse should be designed to meet the specific requirements of the supply chain of which it is a part. In many cases, the need to provide better service to customers and be responsive to their needs appears to be the primary reason. Nevertheless, there are certain operations that are common to most warehouses:

- *Temporarily store goods*. To achieve economies of scale in production, transportation, and handling of goods, it is often necessary to store goods in warehouses and release them to customers as and when the demand occurs.
- *Put together customer orders*. Goods are received from order picking stock in the required quantities and at the required time to the warehouse to meet customer orders. For example, goods can be received from suppliers as whole pallet quantities, but are ordered by customers in less than pallet quantities.
- *Serve as a customer service facility*. In some cases, warehouses ship goods to customers and therefore are in direct contact with them. So, a warehouse can serve as a customer service facility and handle replacement of damaged or faulty goods, conduct market surveys, and even provide after sales service. For example, many Korean electronic goods manufacturers let warehouses handle repair and do after sales service in North America.
- *Protect goods*. Sometimes manufactured goods are stored in warehouses to protect them against theft, fire, floods, and weather elements because warehouses are generally secure and well equipped.
- *Segregate hazardous or contaminated materials*. Safety codes may not allow storage of hazardous materials near the manufacturing plant. Because no manufacturing takes place in a warehouse, this may be an ideal place to segregate and store hazardous and contaminated materials.
- *Perform value-added services*. In many warehouses after picking, goods are brought together and consolidated as completed orders ready to be dispatched to customers. This can involve packing into dispatch outer cases and cartons, and stretch- and shrink- wrapping for load protection and stability, inspecting, and testing. Here, inspection and testing do not add value to the product. However, we have included them because they may be a necessary function because of company policy or federal regulations.
- *Store seasonal inventory*. It is always difficult to forecast product demand accurately in many businesses. Therefore, it may be important to carry inventory and safety stocks to meet unexpected surges in demand. Some companies that produce seasonal products—for example, lawn mowers and snow throwers—may have excess inventory left over at the end of the season and have to store the unsold items in a warehouse.

A typical warehouse consists of two main elements:

1. Storage medium
2. Material handling system

In addition, there is a building that encloses the storage medium, goods, and the S/R system. Because the main purpose of the building is to protect its contents from theft and weather elements, it is made of strong, lightweight material. So, warehouses come in different shapes, sizes, and heights, depending on a number of factors, including the kind of goods stored inside, volume, type of S/R systems used. For example, the Nike warehouse in Laakdal, Belgium, covers a total area of 1 million square feet. Its high-bay storage is almost 100 feet in height, occupies roughly half of the total warehouse space, and is served by 26 man-aboard stacker cranes.

6.3 Inverse Storage

There is limited landfill space available for dumping wastes created throughout the supply chain. And the increasing cost of landfills, environmental laws and regulation, and the economic viability of environmental strategies are pushing manufacturers nowadays to consider reverse supply chain—also known as *reverse logistics*—management.

Manufacturers now must take full responsibility for their products through the product's life cycle, or they may be subject to legal action. For example, new laws regarding the disposal of motor or engine oil, vehicle batteries, and tires place the disposal responsibility on the manufacturer once these products have passed their useful life. Many manufacturers also realize that reverse logistics offers the opportunity to recycle and reuse product components and reduce the cost and the amount of waste. Therefore, manufacturers are developing disposition stocking areas and collecting used or expired original products from the customer and reshipping to their stocking places. For example, Kodak's single-use camera has a remarkable success story involving the inverse logistics philosophy. The products are collected in a stocking place to be remanufactured. In the United States, 63 percent return rate has been achieved for recycling. The details about the procedure can be obtained from Kodak's Web site at http://www.kodak.com/US/en/corp/environment/performance/recycling.html.

7 AVS/RS CASE STUDY

Savoye Logistics is a European logistics company that designs, manufactures, and integrates logistical systems. It provides solutions for order fulfillment and packing and storing/retrieval of unit loads.

Savoye has various teams to assist its customers with logistic expertise to provide them the best solution corresponding to their needs. Their aim is

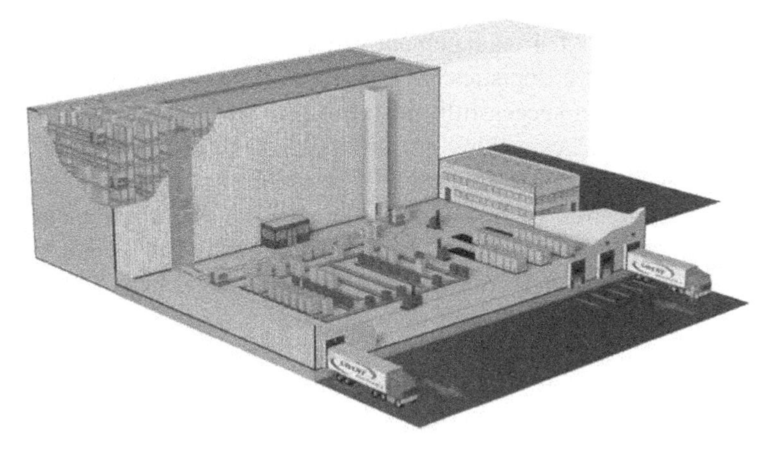

Figure 1.16 Three-dimensional view of a warehouse with an AVS/RS

Figure 1.17 Main components of an AVS/RS (Source: Courtesy of Savoye Logistics)

to guarantee the performance to the customer by selecting the best equipment and a global management of the entire project.

Savoye Logistics has introduced the AVS/RS shown in Figure 1.16. The system has been successfully installed in 35 companies in Europe. Today, the installed systems' capacities are around 1,000,000 pallets and 100,000 movements per day, in eight countries.

Figures 1.17a to 1.17c illustrate the two components of the AVS/RS, one of which is autonomous vehicle and the other is the lift. Although the autonomous vehicle moves horizontally in the storage areas in a given tier, the lift moves the vehicle between tiers. In other words, autonomous vehicles move on rails in the aisles and interface with lifts for vertical movement of pallets between storage tiers. Here, lifts are like conveyors, but they can travel only vertically. Autonomous vehicles also transport pallets between lifts and shipping/receiving areas at the ground level. They can transform the loads from their stored areas to their respective storage addresses in the same tier because they can move within tiers. If the load movement is not on the same tier, then lifts are used for transferring the load to the related tier.

The different load movement patterns make AVS/RSs more flexible than AS/R systems, although at slightly lower efficiency. In AS/RSs, aisle-captive cranes are the main S/R devices to move unit loads simultaneously in the horizontal and vertical dimensions. Unlike storage cranes in AS/RS, AVS/RS vehicles can access any designated storage address but must move in a sequential, rectilinear pattern.

One of Savoye Logistics' AVS/RS applications completed in 1999 was for a telecommunication company that faced rapid growth in one of its warehouses. The logistical challenge in planning was to link the technological manufacturing levels of two buildings with a production supply chain to sustain the material flow from the assembly lines to dispatch. The AVS/RS designed by Savoye was able to satisfy these constructional requirements with a supply and unloading line offset at an angle of 90 degrees. The system has now been in operation and fulfils the short lead times and safety requirements and can achieve fill rate of 95 percent.

REFERENCES

Apple, J. M. 1977. *Plant layout and material handling*, 3rd ed. New York: Wiley.

Ayres, R. U. 1988. Complexity, Reliability, and Design: Manufacturing Implications. *Manufacturing Review* 1: 26–35.

Heragu, S. S. 2008. *Facilities design*, 3rd ed. Clermont, FL: CRC Press.

Lashkari, R. S., R. Boparai, and J. Paulo. 2004. Towards an integrated model of operation allocation and materials handling selection in cellular manufacturing system. *International Journal of Production Economics* 87: 115–139.

Little, J. D. C. 1961. A proof for the queuing formula $L = \lambda W$. *Operations Research* 9: 383–385.

Meyers, F. E. 1993. *Plant layout and material handling*. Englewood Cliffs, NJ: Regents/Prentice Hall.

Paulo, J., R. S. Lashkari, and S. P. Dutta. 2002. Operation allocation and materials handling system selection in a flexible manufacturing system: A sequential modeling approach. *International Journal of Production Research* 40: 7–35.

Sujono, S., and R. S. Lashkari. 2007. A multiobjective model of operation allocation and material handling system selection in FMS design. *International Journal of Production Economics* 105: 116–133.

CHAPTER 2

ERGONOMICS OF MANUAL MATERIALS HANDLING

James L. Smith, Jeffrey C. Woldstad, and Patrick Patterson
Texas Tech University
Lubbock, Texas

1 INTRODUCTION

Ergonomics is the science that matches the capabilities and limitations of humans to the demands of their jobs. The term *jobs* can range from very traditional and literal activities, such as using hand tools or operating equipment, to more figurative jobs such as driving a car or using a cell phone. In the context of materials handling, the job may range from traditional manual materials handling to operating materials handling equipment such as lift trucks or pallet jacks.

In any ergonomics evaluation where potential for injury exists for the workers involved, three strategies are generally proposed. The first level of risk reduction involves *engineering controls*, where the hazardous situation is designed out of the task or is eliminated from the worker's job. An example such as automation, where a machine replaces a human as a materials handler, is a common engineering control. If an engineering solution is not feasible or technologically possible, the next level of risk reduction involves *administrative controls*, where strategies such as limiting exposure through job rotation or the use of multiperson lifting teams are utilized. As opposed to eliminating or reducing the hazard, as is the goal of engineering controls, administrative controls recognize the hazard and attempt to reduce the risk to workers by incorporating strategies such as preemployment strength testing, providing adequate rest recovery, or limiting exposure to the "risky" jobs. As a final defense to reduce worker risk, personal protective equipment (PPE) has been utilized. In the case of manual materials handling, PPE has generally taken the form of *back belts*. PPE is generally regarded by ergonomists as a last resort, or stopgap intervention, until an engineering solution can be found to reduce worker risk.

For many years, workers have been being injured as they lift, lower, push, pull, carry, and otherwise manhandle loads. The jobs in industry vary widely, from warehousing to manufacturing to maintenance to service industries such as parcel delivery, where workers are exposed to the physical demands of handling materials. OSHA (the Occupational Health and Safety Administration) has explored the possibility of establishing manual materials handling standards, but has been unsuccessful to date in promulgating such standards. However, in the absence of such standards, OSHA still has the ability to protect workers exposed to hazardous work conditions through the *General Duty Clause* of the OSHA Act. The General Duty Clause does not allow industry to ignore hazardous work conditions simply because no OSHA standard addresses the specific issue. Despite the presence of OSHA, there are a variety of economic and social issues that would motivate industry to provide a safe work environment for employees and reduce accidents and injuries in the workplace.

2 NIOSH WORK PRACTICES GUIDE

In the 1970s, the National Institute for Occupational Safety and Health (NIOSH, a research arm for OSHA) called together prominent researchers in manual materials handling to address the growing problems associated with manual materials handling injuries. The outcome of that effort was the publication of the *Work Practices Guide for Manual Lifting* in 1981 (NIOSH 1981). The *Work Practices Guide* represented a systematic analysis of manual materials handling research that had been conducted through the 1970s.

2.1 Four Approaches for Studying Materials Handling

The authors recognized four basic approaches that had been utilized in previous manual materials handling research: *epidemiological, biomechanical, physiological*, and *psychophysical*. As they presented the rationales for the four approaches, they developed a recommendation for the analysis of manual lifting tasks. This section presents brief explanations of the four approaches.

2.1.1 Epidemiological Approach

The epidemiological approach focused on the cause–effect relationships that might exist between manual materials handling accidents and injuries and the characteristics of the worker, job, and work environment. The epidemiological approach had received less attention than the other approaches because of the difficulties in establishing the relationships between workers, the work environment, and accidents and injuries. As stated in the 1981 NIOSH guide:

> For injuries to the limbs and those which are superficial, the evidence is usually well documented because the cause and effect are often simple to diagnose. Musculoskeletal injuries (especially to the lower back) are less clear cut and the extent of trauma is seldom defined. The interpretation of such injuries therefore depends mainly on the mechanism of injury and this (due to inexperienced, incomplete or subjective reporting) is difficult to analyze. (NIOSH, 1981)

Although traumatic injuries, such as breaking a leg from a fall while carrying a load, might be relatively simple to analyze, cumulative injuries occurring after repetitive exposure to a particular materials handling situation is much more difficult to analyze. For example, why is it that one person could lift loads repeatedly over days, months, or years and not be injured, while another person performing the same job develops lower back pain after a much shorter exposure time on the job?

The epidemiological approach identifies seven job risk factors that are hazardous to a manual materials handler (NIOSH 1981):

1. *Weight*. What is the force required?
2. *Location/site*. What is the position of the load center of gravity with respect to the worker?
3. *Frequency/duration/pace*. What are the temporal aspects of the task in terms of repetitiveness of handling?
4. *Stability*. Where is the consistency in location of load center of gravity in handling such things as bulky or liquid materials?
5. *Coupling*. What are the texture, handle size and location, shape, color, and so on?
6. *Workplace geometry*. What are the spatial aspects of the task in terms of movement distance, direction, obstacles, postural constraints, and so on?
7. *Environment*. What factors such as temperature, humidity, illumination, noise, vibration, frictional stability of the foot, are involved?

In addition, the epidemiological approach briefly discussed seven personal risk factors (NIOSH 1981):

1. *Gender*. There is no consistent effect, secondary to the strength factor.
2. *Age*. The greatest incidence of low back pain is in the 30- to 50-year-old group.

3. *Anthropometry*. The selection of workers based on anthropometry (especially height or weight) is not justified.
4. *Lift technique*. No controlled epidemiological studies validate any specific lifting techniques.
5. *Attitude*. There is no clear evidence relating attitude to injury risk.
6. *Training*. This is generally accepted as positive, although epidemiological support is lacking.
7. *Strength*. There is epidemiological support that risk is increased when strength capacity is less than job demand.

NIOSH (1981, p. 22) concluded that "heavy load lifting contributes to increased frequency and severity rates for low back pain."

2.1.2 Biomechanical Approach

Tasks that involve relatively heavy loads handled relatively infrequently are generally considered to be biomechanically limiting. The biomechanical approach examines the body as a system of levers (bones) and forces (muscles) and examines the static and dynamic forces on body segments and joints as the worker performs manual materials handling tasks. One of the issues faced in the biomechanical approach involves determining the limitations of the body segments and tissues when subjected to loading.

Probably the system receiving the most attention has been the lower back area, specifically the L5/S1 area and its corresponding disc. Obviously, the only source of data relating to compressive disc failure has been that obtained from cadaver studies. The large variability of disc failures, the validity of cadaver data, and the limited samples have made it difficult to establish biomechanical failure points for the lower back. The NIOSH (1981, p. 36) conclusion regarding biomechanical limitations was, "Jobs that place more than 650 kg compressive force on the low-back are hazardous to all but the healthiest of workers. In terms of a specification for design, a much lower level of 350 kg or lower should be viewed as an upper limit." As a side note, the 1981 NIOSH guide expressed forces in kilograms, when the proper units for forces should have been newtons, a change that was recognized and corrected in later NIOSH publications regarding manual materials handling recommendations.

The 1981 NIOSH guide listed six conclusions for the biomechanics of lifting:

1. Lifting a 5 kg compact load (wherein the mass CG of the load is within 50 cm of the ankles) could create compressive forces sufficient to cause damage to older lumbar vertebral disks.
2. As the load mass center of gravity is moved horizontally away from the body, a proportional increase in the compressive force on the lower back is created. Thus, even light loads need to be handled close to the body.

3. When a load is lifted from the floor, additional stresses are exerted on the lower back due to the body weight moment when stooping to pick up the load. Thus, heavy loads should not be stored on the floor, but should be raised to about standing knuckle height (minimum 50 cm) to avoid the necessity of stooping over and lifting.
4. The postures used to lift loads from the floor can exert a complex and relatively unknown effect on the stressors of the lower back during lifting. Specific instructions as to the safe lifting posture to use will be necessarily complex, reflecting such factors as leg strengths, load, and load size. Until such complexities are better researched, it is recommended that instructions as to lifting postures be avoided.
5. Lifting loads asymmetrically (by one hand or at the side with the torso twisted) can impart complex and potentially hazardous stresses to the lumbar column. Such acts should be avoided by instructions and workplace layouts, which permit the worker to address the load in a symmetric manner.
6. The dynamic forces imparted by rapid jerking motions can multiply a load's effect greatly. Instructions to handle even moderate loads in a smooth and deliberate manner are recommended.

2.1.3 Physiological Approach

Tasks that involve relatively lighter loads handled relatively frequently are generally considered to be physiologically limiting. The physiological approach generally utilizes energy expenditure, oxygen consumption, or heart rate to describe the demands of the materials handling task. However, when examining physiological fatigue, both static as well as dynamic tasks must be considered. Static (or isometric) tasks might be considered to be infrequently occurring in manual materials handling until it is recognized that activities such as carrying, holding, or maintaining fixed postures have significant static components to them. For static tasks, endurance limitations can be as low as a few (3 to 5) seconds for efforts demanding 100 percent of a worker's maximum voluntary contraction (MVC) or strength, to very long endurance times for tasks demanding less than 15 percent of MVC.

For dynamic activities where relatively large muscle masses are utilized, physiological demand can be represented by energy expenditure, oxygen consumption, or heart rate. Generally, a linear relationship has been assumed between heart rate and oxygen uptake (VO_2). Assuming that maximum heart rate can be approximated by $HR_{max} = 220 - \text{Age}$ (in years), a worker's physiological capacity ($VO_{2\,max}$) can be determined through submaximal testing and determining the linear relationship between heart rate (HR) and oxygen consumption (VO_2). Techniques for determining $VO_{2\,max}$ can be found in references such as Tayyari and Smith (1997). A conclusion of the physiological approach in the NIOSH (1981) guide was that "33 percent of aerobic capacity will be assumed for 8-hour work duration."

The 1981 NIOSH guide arrived at the following physiological design criteria:

1. For occasional lifting (for one hour or less) metabolic energy expenditure rates should not exceed 9 Kcal/min for physically fit males or 6.5 Kcal/min for physically fit females.
2. Likewise, continuous (8-hour) limits should not exceed 33 percent of aerobic capacity or 5 Kcal/min and 3 Kcal/min, respectively. These guideline limits do not reflect the increased metabolic rates, which would be associated with overweight or deconditioned workers.
3. Personal attributes of age, gender, body weight, and so on are insufficient to accurately predict work capacity for any particular individual, although such data are sufficient for making predictions of group averages.
4. The primary task variables which influence metabolic rate during lifting are: (a) load handled, (b) vertical location at beginning of lift, (c) vertical travel distance, and (d) frequency of lift.

2.1.4 Psychophysical Approach

Psychophysics is based on a research methodology that states that the body integrates stresses (physical, mental, and environmental) and responds with an output that reflects the worker's "acceptable level of response to the stress." Typically, worker capacity data are collected through experimentation in which the worker determines his or her maximum acceptable response to a set of task conditions. For example, in determining the psychophysical lifting capacity of an individual, that individual would adjust weights in a container until the maximum amount of weight that could be handled under the given task conditions is determined. The procedure attempts to eliminate factors such as visual clues by placing false bottoms in the container, with various amounts of weights hidden in the false bottom to eliminate a visual clue as to how much weight the person was selecting. Also, individuals are given no feedback regarding their weight selection or performance and are tested in isolation in an attempt to eliminate competitive factors. If conducted properly, researchers can usually expect people to give responses within 10 percent variability from day to day. Psychophysical responses are often summarized as levels acceptable to various percentages of the population. For example, a 30-pound load lifted once a minute from floor to knuckle height might be determined to be acceptable to 25 percent of the male population and less than 5 percent of the female population. An excellent summary of psychophysical materials handling capabilities can be found in Snook (1978) and Ayoub et al. (1978).

Some of the psychophysical approach conclusions of the 1981 NIOSH guide include these three:

1. The majority of lower back injuries were shown to occur on jobs that were not "acceptable" to 75 percent of the population. For a female workforce, this percentile specification should be protective; however, for a male workforce, the limit would be overly restrictive.
2. For low-frequency lifting, capacities are limited by strength rather than endurance.
3. Psychophysical criteria are most appropriate for occasional moderate to higher frequency lifting.

2.2 NIOSH 1981 Recommendations

The committee arrived at recommendations for evaluating manual materials handling tasks and published those recommendations in the *Work Practices Guide for Manual Lifting* in 1981. The recommendations were based on the following set of assumptions or limitations and apply only to lifting tasks:

1. Smooth lifting (no jerking or erratic patterns)
2. Two-handed, symmetric lifting in the sagittal plane (directly in front of the body; no twisting during lift)
3. Moderate width [e.g., 70 cm (30 inches) or less]
4. Unrestricted lifting posture
5. Good couplings (handles, shoes, floor surfaces)
6. Favorable ambient conditions

It was further stated that the guide does not include any "safety factors" commonly used by engineers to assure that unpredicted conditions are accommodated.

The recommendations of the 1981 guide resulted in the establishment (calculation) of two limits: an *action limit* (AL) and a *maximum permissible limit* (MPL). For tasks with demands below the AL, those tasks should represent minimal risk to workers and should be monitored but not require further action. For those tasks where task demand exceeded the MPL, significant risks are present and the tasks should be redesigned to reduce task demand to a more acceptable level (engineering controls were recommended). Task demands falling between the AL and MPL indicate some risk and should be modified to be more acceptable (administrative or engineering controls recommended).

The four research approaches were related to the AL and MPL as follows (NIOSH 1981):

At the Maximum Permissible Limit (MPL)

- *Epidemiological*. Musculoskeletal injury incidence and severity rates have been shown to increase significantly in populations when work is performed above the MPL.

- *Biomechanical*. Compressive forces on the L5/S1 disc are not tolerable over 650 kg (1,430 lbs) in most workers, which would result in conditions above the MPL.
- *Physiological*. Metabolic rates would exceed 5.0 kcal/min for most individuals working above the MPL.
- *Psychophysical*. Only about 25 percent of men and less than 1 percent of women workers have the muscle strengths to be capable of performing work above the MPL.

At the Action Limit (AL)

- *Epidemiological*. Musculoskeletal injury incidence and severity rates increase moderately in populations exposed to lifting conditions described by the AL.
- *Biomechanical*. A 350 kg (770 lbs) compression force on the L5/S1 disc can be tolerated by most young, healthy workers, which would result in conditions at the AL.
- *Physiological*. Metabolic rates would exceed 3.5 Kcal/min for most individuals working above the AL.
- *Psychophysical*. Over 75 percent of women and over 99 percent of men could lift loads described by the AL.

The AL and MPL are calculated as follows:

$$\begin{aligned} \text{AL (kg)} &= 40\,(15/H)(1 - 0.004\,|V - 75|)(0.7 + 7.5/D)(1 - F/F_{\max}) \\ \text{AL (lb)} &= 90\,(6/H)(1 - 0.01\,|V - 30|)(0.7 + 3/D)(1 - F/F_{\max}) \end{aligned} \tag{1}$$

$$MPL = 3(\text{AL}) \tag{2}$$

where:

H = Horizontal location (centimeters or inches) forward of midpoint between the ankles at the origin of lift
V = Vertical location (centimeters or inches) at the origin of lift
D = Vertical travel distance (centimeters or inches) between the origin and destination of lift
F = Average frequency of lift (lifts/minute)
$F_{\max}$ = Maximum frequency, which can be sustained (see Table 2.1)

Table 2.1 $F_{\max}$ Table

	$V > 75$ (30) Standing	$V \leq 75$ (30 in) Stooped
1 hour duration	18	15
8 hours duration	15	12

The variables in equations (1) and (2) were subject to the following limits (NIOSH 1981):

H is between 15 cm (6 in.) and 80 cm (32 in.). Objects cannot, in general, be closer than 15 cm (6 in.) without interference with the body. Objects further than 80 cm (32 in.) cannot be reached by many people.

V is assumed between 0 cm and 175 cm (70 in.), representing the range of vertical reach for most people.

D is assumed between 25 cm (10 in.) and $(200 - V)$ cm $[(80 - V)$ in.]. For travel less than 25 cm, set $D = 25$.

F is assumed between 0.2 (one lift every 5 minutes) and F_{max} (see Table 2.1). For lifting less frequently than once every 5 minutes, set $F = 0$.

Most common errors in calculating the AL from the 1981 NIOSH guide arose from not using the limitations for each of the factors. One way to self-check for calculation errors is to consider the four factors in the equation (the calculations in parentheses) as modifiers to an initial load of 40 kg or 90 lbs. The factors can only have values of 1 or less. If a factor has a value greater than 1, there has been an error (most likely one of the limitations was not followed). For example, suppose a load is lifted only 4 inches onto a platform. The vertical travel distance factor would be $(0.7 + 3/D)$. If $D = 4$ were substituted into the equation, the value of the factor would be $(0.7 + 3/4)$, or 1.45. Since this value exceeds 1.0, there has been an error. In checking the limits, it is seen that for D less than 10 inches, set $D = 10$. Now, going back to our vertical travel distance factor, the proper calculation would be $(0.7 + 3/10)$, or 1.0.

For the more graphical individuals, NIOSH (1981) presented the four factors in graphical form in nomograms. One advantage of the nomograms is that no factor can be greater than 1.0 when using all four nonograms (see Figures 2.1–2.4). In

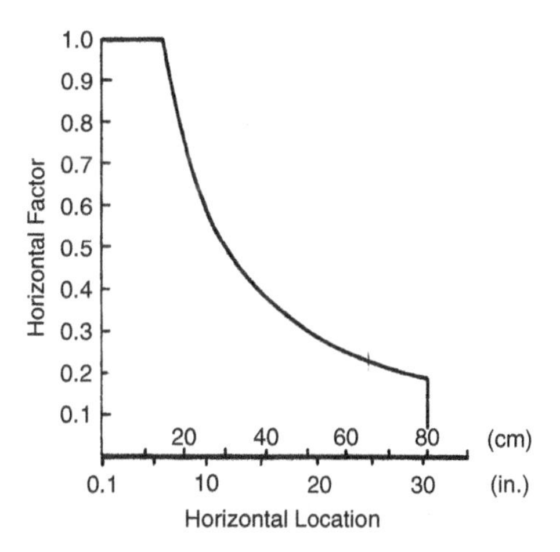

Figure 2.1 Horizontal factor graph

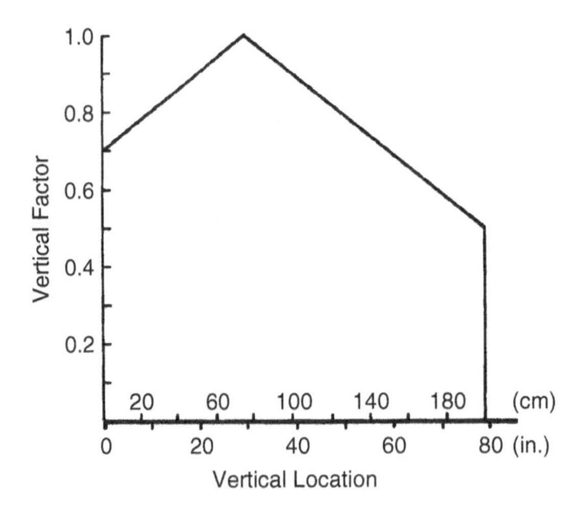

Figure 2.2 Vertical factor graph

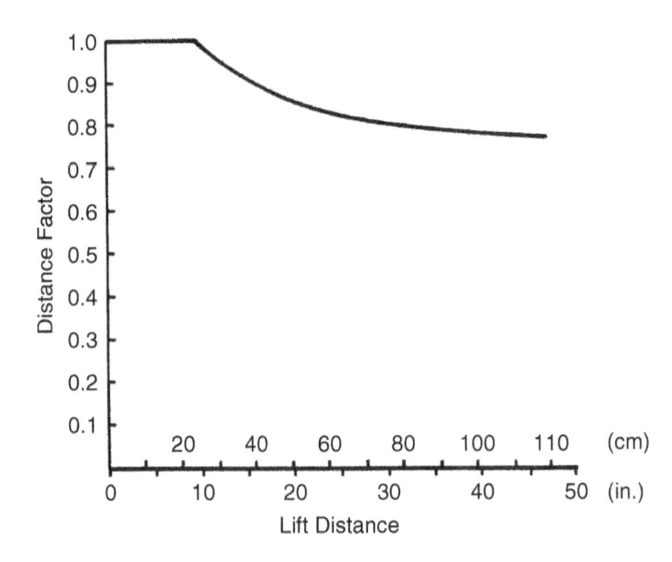

Figure 2.3 Vertical travel distance factor

its simplest form, the calculations for AL and MPL using the nomograms can be found as follows:

$$\begin{aligned} \text{AL (kg)} &= 40\ (\text{Horizontal factor})\ (\text{Vertical factor})\ (\text{Distance factor}) \\ &\quad (\text{Frequency factor}) \\ \text{AL (lb)} &= 90\ (\text{Horizontal factor})\ (\text{Vertical factor})\ (\text{Distance factor}) \\ &\quad (\text{Frequency factor}) \end{aligned} \tag{3}$$

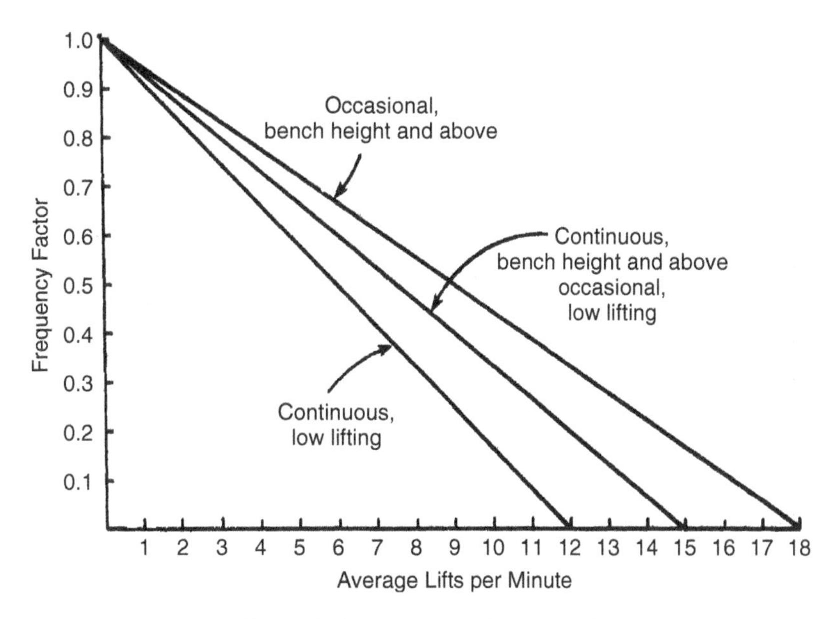

Figure 2.4 Frequency factor graph

$$\text{MPL} = 3(\text{AL}) \tag{4}$$

See Figures 2.1, 2.2, 2.3, and 2.4.

2.3 Example Using 1981 NIOSH Guide

For two hours per day, an employee lifts objects off a feed conveyor and places them in a packing carton after a visual inspection. The objects are 25 cm cubes and arrive at a rate of four per minute. The conveyor is 70 cm high and the bottom of the packing station is 80 cm from the floor. Assume that the objects on the conveyor are very close to the person lifting them (assume that H is $15 + W/2$).

Identify the variables:

Load constant = 40 kg

$H = 15 + W/2 = 15 + 25/2 = 27.5$ cm

$V = 70$ cm

$D = (80 - 70) = 10$

However, based on the limitation for D, $D = 25$ cm for the calculation:

$F = 4$ lifts/minute

$F_{max} = 12$ lifts/min

$\text{AL (kg)} = 40\,(15/27.5)(1 - 0.004\,|70 - 75|)(0.7 + 7.5/25)(1 - 4/12)$

$\text{AL (kg)} = 40\,(0.545)(0.98)(1.0)(0.667)$

AL = 14.25 kg
MPL = 42.75 kg

If the object mass were 10 kg, then the load would be less than the *AL* and consequently no further consideration is required—the task represents minimal risk to the workers.

If the object mass were 20 kg, then the task would represent more than minimal risk and require attention. In this case, administrative or engineering controls would be suggested to increase the AL to a value greater than 20 kg. In modifying the task, the factors in the AL calculation can provide valuable insights as to which modifications will provide the greatest benefit. In this case, increasing the height of the feed conveyor to 75 cm would yield only a 2 percent gain in AL, which is insufficient. The strategy should be to examine the lowest factors to see if they can be increased to yield the greatest increase in the AL. In this case, the horizontal factor of 0.545 is the greatest limitation, followed by the frequency factor at 0.667. Can the box size be reduced to get the load closer to the body? If not, then the frequency factor must be examined. In this case, limiting exposure of the employee to one hour will result in about a 10 percent increase in the AL, which is still not satisfactory. Raising the feed conveyor will result in about a 19 percent increase in AL (2 percent from vertical factor and 17 percent from frequency factor). This will still leave the AL below 20. Therefore, in this case the horizontal factor is the only factor that will allow the AL to become greater than 20 (assuming that reducing the frequency to 1/minute would not be acceptable). Therefore, engineering controls would be the appropriate answer to this problem. Obviously, if the load were greater than 42.75 kg, it would be above the MPL and engineering controls would also be required. Potential engineering controls might include automation (automated packing, robotic lifting, mechanical lift assists, etc.) or repackaging of the product into small, lighter packages, or possibly repackaging into larger unit loads that would be handled with automated equipment (sometimes referred to as the unit load principle in materials handling).

The 1981 NIOSH lifting guide represented a widely used method for assessment of manual materials handling. One of the primary concerns in using the AL/MPL calculations involved the assumptions used for the equations development. How many tasks in industry were smooth, two-handed sagittal plane lifting of moderate width packages with handles under favorable ambient conditions? Despite the limitations, the lifting guide gave ergonomists a starting point for analysis of lifting tasks. Recognizing the widespread availability of the *NIOSH Work Practices Guide*, OSHA began issuing citations (under the General Duty Clause) for situations where workers were being exposed to high-risk manual materials handling tasks, and companies were taking no action to reduce those injuries.

2.4 Revised NIOSH Guide

In the late 1980s, another committee was formed by NIOSH (several members were also members of the original committee) to address the shortcomings of the 1981 guide and to explore revisions to the guide. The result of the committee's efforts was the publication of the "Revised NIOSH Equation for the Design and Evaluation of Manual Lifting Tasks" (Waters et al 1993) and *Applications Manual for the Revised NIOSH Lifting Equation* (Waters et al. 1994). The major changes of the revised guide included adding factors for couplings (handles) and accounting for asymmetry during lifting. The concepts of the action limit (AL) and the maximum permissible limit (MPL) were replaced with a recommended weight limit (RWL) and a lifting index (LI). The RWL is similar to the AL calculation in that it consists of a load constant that is then modified by six adjustment factors: horizontal factor, vertical factor, distance factor, frequency factor, coupling factor, and asymmetry factor. The LI is a ratio of task demand (load weight) to the RWL. With the inclusion of asymmetry, one of the original assumptions of the 1981 guide of symmetric sagittal plane lifting was removed.

From three of the approaches outlined in the 1981 guide, the criteria in Table 2.2 were used to develop the revised guide lifting equations (Waters et al. 1993). When the criteria for the revised guide are compared to the AL criteria from the 1981 guide, similarities can be noted (see Table 2.3).

As can be seen in the tables, the biomechanical criteria remained the same (although the more proper force units of newtons were used), the 1981 physiological criteria is the midpoint of the range found in the revised guide, and the psychophysical criteria remained unchanged. The revised guide lowered the load constant but then decreased the adjustments due to the horizontal factor, the

Table 2.2 Criteria for Revised Lifting Guide

Approach (revised)	Design Criteria	Cut-off Value
Biomechanical	Max disc compression force	3.4 kN (770 lbs)
Physiological	Max energy expenditure	2.2 to 4.7 kcal/min
Psychophysical	Max acceptable weight	Acceptable to 75% of female workers and 99% of male workers

Table 2.3 Criteria for Original Lifting Guide

Approach (1981)	Design Criteria	Cut-off Value
Biomechanical	Max disc compression force	350 kg (770 lbs)
Physiological	Max energy expenditure	3.5 kcal/min
Psychophysical	Max acceptable weight	Acceptable to 75% of female workers and 99% of male workers

vertical factor, and the distance factor, while increasing the effect of frequency and adding reduction factors for couplings and asymmetry.

2.4.1 *Revised Lifting Equations*

The revised lifting equations are as follows (Waters et al. 1994):

$$\begin{aligned} \text{RWL} &= \text{LC} \times \text{HM} \times \text{VM} \times \text{DM} \times \text{AM} \times \text{FM} \times \text{CM} \\ \text{LI} &= (\text{Load weight})/(\text{RWL}) \end{aligned} \tag{5}$$

Table 2.4 lists the metrics for the variables in equation (5).

2.4.2 *Definition and Constraints of the Factors in the Revised Guide*

As in the 1981 guide, factors (multipliers) are defined and limitations are presented for each factor (Waters et al. 1994):

- *Horizontal location*. Measured from the midpoint of the line joining the inner ankle bones to a point projected on the floor directly below the midpoint of the hand grasps (i.e., load center) as defined by the large middle knuckle of the hand. If significant control is required at the destination, then H should be measured at both the origin and destination.

  $\text{HM} = (25/H)$ in cm, or

  $\text{HM} = (10/H)$ in inches

  When H cannot be measured, it can be approximated using Table 2.5.

  If $H \leq 25$ cm (10 in.), then set $H = 25$ cm (10 in.), and the horizontal multiplier becomes 1.0. If $H > 63$ cm (25 in.), then the horizontal multiplier becomes 0.0.

Table 2.4 Factors for Revised Lifting Equations

		Metric	U.S. Customary
Load Constant	LC	23 kg	51 lb
Horizontal Multiplier	HM	(25/H)	(10/H)
Vertical Multiplier	VM	1 − (0.003 |V − 75|)	1 − (0.0075 |V − 30|)
Distance Multiplier	DM	0.82 + (4.5/D)	0.82 + (1.8/D)
Asymmetric Multiplier	AM	1 − (0.0032 A)	1 − (0.0032 A)
Frequency Multiplier	FM	From Table 2.11	From Table 2.11
Coupling Multiplier	CM	From Table 2.12	From Table 2.12

Table 2.5 Horizontal Factor Calculations (H Cannot Be Measured)

Metric (Distances in cm)	U.S. Customary (Distances in inches)
$H = 20 + W/2$ For $V \geq 25$ cm	$H = 8 + W/2$ For $V \geq 10$ in.
$H = 25 + W/2$ For $V < 25$ cm	$H = 10 + W/2$ For $V < 10$ in.

- *Vertical location*. Vertical height of the hands above the floor (at origin of lift). V is measured vertically from the floor to the mid-point between the hand grasps, as defined by the large middle knuckle.

  $\mathrm{VM} = (1 - 0.003\,|V - 75|)$ in cm, or

  $\mathrm{VM} = (1 - 0.0075\,|V - 30|)$ in inches

  If $V > 175$ cm (70 in.), then the vertical multiplier becomes 0.0. It is assumed that V cannot be less than zero (not lifting a load below floor level).

- *Distance component*. Vertical travel distance of the hands between the origin of the lift and destination of lift.

  $\mathrm{DM} = (0.82 + 4.5/D)$ in cm

  $\mathrm{DM} = (0.82 + 1.8/D)$ in inches

  D is assumed to be at least 25 cm (10 in.) and no greater than 175 cm (70 in.). If $D < 25$ cm ($<$ 10 in.), then use $D = 25$ cm or $D = 10$ inches in the equations just given.

- *Asymmetry component*. Refers to a lift that begins or ends outside the midsagittal plane of the body, which is the plane going through the middle of the body separating the body into left and right sides. The AM decreases about 0.14 for each 45 degrees of asymmetry of lift.

  $\mathrm{AM} = 1 - 0.0032A$, where A is the angle of asymmetry

  The angle A is limited to the range of 0 to 135 degrees. If $A > 135$ degrees, the AM = 0 and RWL = 0.

- *Frequency component*. This is defined by the number of lifts per minute, the task duration, and the height of the origin of lift. Lifting frequency should be observed and measured over at least a 15-minute period. Lift duration has been defined in three time intervals rather than the two intervals of the 1981 guide. The durations are as follows:
  - *Short*. One hour or less, followed by a recovery time period of 1.2 times the work time (e.g., 30 minutes work, followed by at least 36 minutes of recovery would be considered short duration).
  - *Moderate*. More than one hour but no more than two hours, followed by a recovery period of at least 0.3 times the work time (e.g., 100 minutes of work, followed by 30 minutes of recovery).
  - *Long*. More than two hours work up to eight hours. The height of the origin of lift differentiates between lifts above or below knuckle height (75 cm or 30 in.).

  For frequencies of less than one lift every five minutes ($F \leq 0.2$), use $F = 0.2$. In the 1981 guide, F_{max} values were 12, 15, or 18 lifts/min, depending on origin of lift and duration of lifting. In the revised guide, F_{max} values as low as 8 lifts/min are used for long-duration, low (below knuckle height) origins of lift.

A special procedure was developed for situations in which the worker did not lift continuously during the 15-minute observation period. For example, if a worker lifts at a rate of 10 lifts per minute for 8 minutes, followed by 7 minutes of light work, the frequency F would be calculated as $F = 10(8)/15 = 5.33$ lifts/min. The analyst should be careful to ensure that the 15-minute observation period was representative of the total task demand of the worker.

- *Coupling component*. The nature of the coupling, as well as the location of the origin of lift, were used to determine the coupling multiplier (CM). The coupling multiplier ranges from 0.90 to 1.0. Table 2.6 is used to classify the nature of the couplings.

2.4.3 *Assumptions of the Revised Guide*

Waters et al. (1993) described five limitations of the revised lifting equation:

1. It assumes that manual materials handling activities other than lifting are minimal and do not require significant energy expenditure.
2. It assumes no unpredicted conditions such as unexpected heavy loads or slips or falls.
3. It assumes no one-handed lifting or lifting while seated or kneeling, lifting in a confined workspace, shoveling, lifting people, lifting hot or cold loads, or high-speed lifting tasks.
4. It assumes worker shoe/floor static coefficient of friction of 0.4 (preferably > 0.5).
5. It assumes that lifting and lowering tasks have the same level of risk.

2.4.4 *Using the Revised Guide*

In using the revised guide, the RWL equation was simplified somewhat by emphasizing the modifiers rather than presenting a numerical equation, as was done to calculate the AL of the 1981 guide. Even though the 1981 guide equation was relatively simple mathematically, many practitioners had difficulty properly calculating the AL. Therefore, the RWL was presented more simply as: RWL = LC (HM) (VM) (DM) (AM) (FM) (CM). The multiplier factors could be calculated or looked up in a provided table, but NIOSH felt that inclusion of a set of six tables (one for each of the six multiplier factors) would reduce the errors in calculating the RWL. The primary advantage of using the tables to find the six modifier factors is that the table values range from only 0.0 to 1.0, so factors greater than 1.0 are eliminated and there is no possibility for a negative factor entering the calculation.

Table 2.6 Classification of Couplings

Good	Fair	Poor
1. For containers of optimal design, such as boxes, crates, and so on, a "good" hand-to-object coupling would be defined as handles, or hand-hold cut-outs of optimal design (see notes 1 to 3).	**1.** For containers of optimal design, a "fair" hand-to-object coupling would be defined as handles or handhold cutouts of less than optimal design (see notes 1 to 4).	**1.** Containers have less than optimal design, or loose parts or irregular objects that are bulky, hard to handle, or have sharp edges (see note 5).
2. For loose parts or irregular objects, which are not usually containerized, such as castings, stock, and supply materials, a "good" hand-to-object coupling would be defined as a comfortable grip in which the hand can be easily wrapped around the object (see note 6).	**2.** For containers of optimal design with no handles or handhold cutouts or for loose parts or irregular objects, a "fair" hand-to-object coupling is defined as a grip in which the hand can be flexed about 90 degrees (see note 4).	**2.** Nonrigid bags (i.e., bags that sag in the middle) would be considered poor containers.

Notes:

1. An optimal handle design has 0.75–1.5 inches (1.9–3.8 cm) diameter, $\geq$ 4.5 inches (11.5 cm) length, 2 inches (5 cm) clearance, cylindrical shape, and a smooth, nonslip surface.
2. An optimal hand-hold cut-out has the following approximate characteristics: $\geq$ 1.5 inch (3.8 cm) height, 4.5 inch (11.5 cm) length, semi-oval shape $\geq$ 2 inches (5 cm) clearance, smooth nonslip surface, and $\geq$ 0.25 inches (0.60 cm) container thickness (e.g., double thickness cardboard).
3. An optimal container design has $\leq$ 16 inches (40 cm) frontal length, $\leq$ 12 inches (30 cm) height, and a smooth nonslip surface.
4. A worker should be capable of clamping fingers at nearly 90 degrees under the container, such as required when lifting a cardboard box.
5. A container is considered less than optimal if it has a frontal length > 16 inches (40 cm), height > 12 inches (30 cm), rough or slippery surfaces, sharp edges, asymmetric center of mass, unstable contents, or requires the use of gloves. A loose object is considered bulky if the load cannot easily be balanced between the hand-grasps.
6. A worker should be able to comfortably wrap the hand around an object without causing excessive wrist deviations or awkward postures, and the grip should not require excessive force.

2.4.5 Example Using the Revised Guide

Use the same example as was used previously for the 1981 guide:

For two hours per day, an employee lifts objects off a feed conveyor and places them in a packing carton after a visual inspection. The objects are 25 cm cubes and arrive at a rate of 4 per minute. The conveyor is 70 cm high and the bottom of the packing station is 80 cm from the floor. Assume that the objects on the conveyor are very close to the person lifting them (assume that H is

$20 + W/2$). For the revised example, we will also assume that the worker is now required to twist 90 degrees when placing the object on the packing station. We will also assume the object has good couplings.

$$\text{RWL} = \text{LC (HM) (VM) (DM) (AM) (FM) (CM)}$$

The LC (load constant) has been defined as 23 kg, or 51 lb.

Tables 2.7 to 2.12 are multiplier tables for the revised lifting equation (see Waters et al. 1994), and can be used to evaluate these equations:

$$H = 20 + 25/2 = 32.5\,\text{cm};\ \text{HM} = 0.78$$

$$V = 70;\ \text{VM} = 0.99$$

$$D = 5\,\text{cm};\ \text{for } D < 25\,\text{cm},\ \text{set } D = 25;\ \text{DM} = 1.00$$

$$A = 90^\circ;\ \text{AM} = 0.71$$

$$F = 4\,\text{lifts/min};\ 1\ \text{hour} \leq \text{duration} \leq 2\ \text{hours};\ \text{FM} = 0.72$$

Coupling is "good"; CM = 1.0

Therefore, the RWL $= 23(0.78)(0.99)(1.00)(0.71)(0.72)(1.00) = 9.08$ kg

Recall from the example using the 1981 guide that AL was calculated to be 14.25 kg. If we were to discount the asymmetry factor, the RWL would be 12.8 kg, which is close to the AL value of 14.3 kg.

Table 2.7 Horizontal Multiplier

H (in.)	HM	*H* (cm)	HM
<10	1.00	<25	1.00
11	0.91	28	0.89
12	0.83	30	0.83
13	0.77	32	0.78
14	0.71	34	0.74
15	0.67	36	0.69
16	0.63	38	0.66
17	0.59	40	0.63
18	0.56	42	0.60
19	0.53	44	0.57
20	0.50	46	0.54
21	0.48	48	0.52
22	0.46	50	0.50
23	0.44	52	0.48
24	0.42	54	0.46
25	0.40	56	0.45
>25	0.00	58	0.43
		60	0.42
		63	0.40
		>63	0.00

Table 2.8 Vertical Multiplier

V (in.)	VM	*V* (cm)	VM
0	0.78	0	0.78
5	0.81	10	0.81
10	0.85	20	0.84
15	0.89	30	0.87
20	0.93	40	0.90
25	0.96	50	0.93
30	1.00	60	0.96
35	0.96	70	0.99
40	0.93	80	0.99
45	0.89	90	0.96
50	0.85	100	0.93
55	0.81	110	0.90
60	0.78	120	0.87
65	0.74	130	0.84
70	0.70	140	0.81
>70	0.00	150	0.78
		160	0.75
		170	0.72
		175	0.70
		>175	0.00

Table 2.9 Distance Multiplier

D (in.)	DM	*D* (cm)	DM
<10	1.00	<25	1.00
15	0.94	40	0.93
20	0.91	55	0.90
25	0.89	70	0.88
30	0.88	85	0.87
35	0.87	100	0.87
40	0.87	115	0.86
45	0.86	130	0.86
50	0.86	145	0.85
55	0.85	160	0.85
60	0.85	175	0.85
70	0.85	>175	0.00
>70	0.00		

The 1981 lifting guide calculated an AL and a MPL value. Task demands below the AL were considered to represent minimal risk, those between the AL and MPL were moderate risk and required engineering or administrative controls, and those above the MPL were high risk and required engineering controls. In the revised lifting guide, task demands below the RWL were deemed to be of

Table 2.10 Asymmetric Multiplier

A deg	AM
0	1.00
15	0.95
30	0.90
45	0.86
60	0.81
75	0.76
90	0.71
105	0.66
120	0.62
135	0.57
>135	0.00

minimal risk and required no further analysis (although continuous monitoring would be recommended in case task conditions changed). The revised NIOSH guide (Waters et al. 1994, p. 35) stated that job risk increased as the RWL increased, and that "nearly all workers will be at an increased risk of work-related injury when performing highly stressful lifting tasks (i.e., lifting tasks that would exceed a LI of 3.0)." Since the MPL was calculated to be three times the AL and the LI range of 1 to 3, the parallels can be seen between the original and the revised lifting guides.

To date, NIOSH has not presented any further revisions to the revised lifting guide (Waters et. al. 1994). Therefore, the revised guide represents the best analytical tool for the ergonomic evaluation of manual materials handling tasks. The *Applications Manual for the Revised NIOSH Lifting Equation* (Waters et al. 1994) goes into significantly more detail in the background and calculations for the revised guide and should serve as a reference to the ergonomist conducting manual materials handling analyses. The manual provides numerous examples and provides details regarding jobs involving multiple lifting tasks (i.e., manual palletizing or depalletizing, where the objects being handled are stacked on multiple layers, so the *H* and *V* values change for each row and layer).

Commercial software packages are available that can perform the calculations. However, as with any software package, the analyst should validate and verify the software before making recommendations based on the software analysis.

2.4.6 *What about Nonlifting Manual Materials Handling Tasks?*

Lifting is the most prevalent manual materials handling task. As such, it has received the most attention by NIOSH and researchers. However, there are many additional materials handling activities that can lead to worker injuries. NIOSH incorporated *lowering* into its analysis in the revised lifting equation by simply stating that lowering activities could be analyzed as lifting activities, which is a

Table 2.11 Frequency Multiplier

FLifts/min	Duration					
	< 1 hour		1–2 hours		2–8 hours	
	V < 30 in	V > 30 in	V < 30 in	V > 30 in	V < 30 in	V > 30 in
<0.2	1.00	1.00	0.95	0.95	0.85	0.85
0.5	0.97	0.97	0.92	0.92	0.81	0.81
1	0.94	0.94	0.88	0.88	0.75	0.75
2	0.91	0.91	0.84	0.84	0.65	0.65
3	0.88	0.88	0.79	0.79	0.55	0.55
4	0.84	0.84	0.72	0.72	0.45	0.45
5	0.80	0.80	0.60	0.60	0.35	0.35
6	0.75	0.75	0.50	0.50	0.27	0.27
7	0.70	0.70	0.42	0.42	0.22	0.22
8	0.60	0.60	0.35	0.35	0.18	0.18
9	0.52	0.52	0.30	0.30	0.00	0.15
10	0.45	0.45	0.26	0.26	0.00	0.13
11	0.41	0.41	0.00	0.23	0.00	0.00
12	0.37	0.37	0.00	0.21	0.00	0.00
13	0.00	0.34	0.00	0.00	0.00	0.00
14	0.00	0.31	0.00	0.00	0.00	0.00
15	0.00	0.28	0.00	0.00	0.00	0.00
>15	0.00	0.00	0.00	0.00	0.00	0.00

Table 2.12 Coupling Multiplier

Coupling Type	CM	
	V < 30 in.	V > 30 in.
Good	1.00	1.00
Fair	0.95	1.00
Poor	0.90	0.90

conservative approach. Other manual handling activities—such as carrying, holding, pushing, pulling, one-handed exertions, and materials handling in unusual postures or under harsh environmental conditions—have yet to be addressed by NIOSH. However, researchers have investigated many of these non-NIOSH lifting activities and have reported their results in the literature. Researchers at Liberty Mutual have utilized the psychophysical approach to examine maximum acceptable push, pull, and carry loads (Snook and Ciriello 1991). Tables of psychophysically determined capabilities for materials handling in unusual postures can be found in Smith et al. (1989 and 1992). The most comprehensive summary of nonlifting task capabilities can be found in *A Guide to Manual Materials Handling*, second edition (Mital et al. 1997). Although there are no NIOSH guides for these "other" manual materials handling activities, the ergonomist is responsible to use the available information to design or redesign workplaces to insure a safe work environment for employees.

3 MECHANICAL MATERIALS HANDLING

Ergonomists have focused on reducing injuries to workers who perform manual materials handling tasks. After utilizing the appropriate tools, such as the NIOSH guidelines, the engineering and administrative control recommendations may represent the best solution to protecting the worker. However, the human impacts of mechanical materials handling devices should be examined before such devices are recommended.

3.1 Device Types and Uses

When an ergonomic analysis of a manual materials handling job (such as the NIOSH *Work Practices Guide*) indicates that tasks components are too strenuous to be safely performed, the use of an automated or semi-automated mechanical material handling device is often prescribed. Automated devices require very little human interaction and are not considered in this section. Semiautomated devices or materials handling assist devices replace a component of the materials handling task, while requiring the worker to manually perform other components of the task. This section describes common material handling assist devices.

3.1.1 Hoists

Hoists are devices used to lift a load and transport it within a prescribed geometric space. Hoist types include triaxial manipulators (Figure 2.5), balancing hoists, jib cranes, and overhead trolley suspension hoists. These devices include a mechanical linkage that supports the load within the prescribed space, some end-of-device tooling that connects the load to the device, and some type of control system for the operator. Although some devices provide assistance in moving the load, most only lift the object and require the operator to push or pull it to the desired

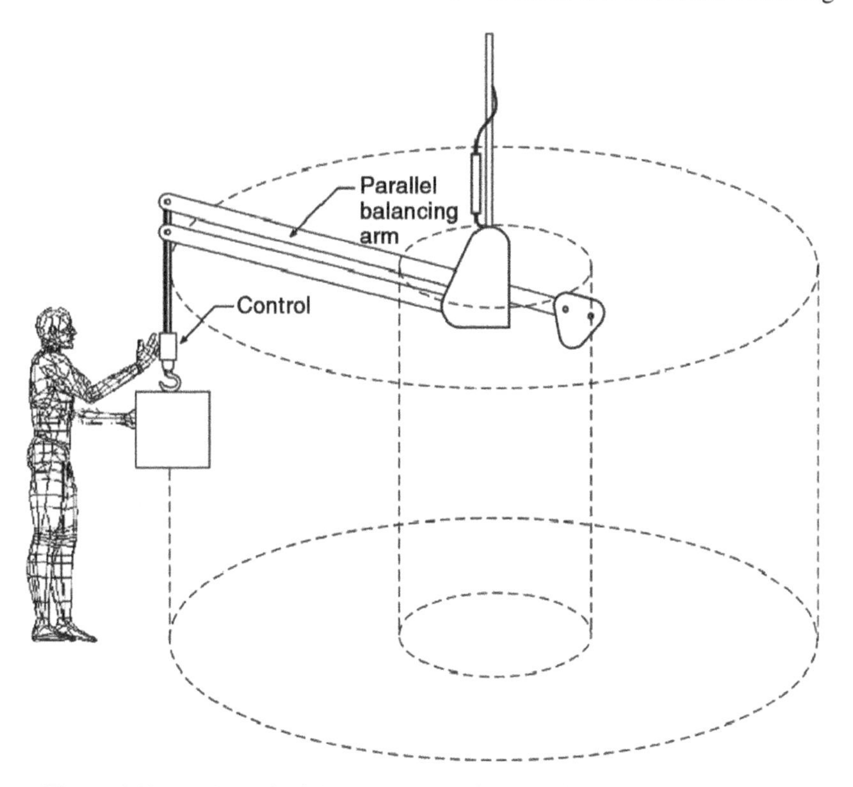

Figure 2.5 Typical triaxial manipulator (from Woldstad and Reasor 1996)

location. Hoists are a good solution when a load must be lifted and moved within a fixed space. They are not good solutions when the load must be transported long distances or when the work area is crowded with other devices or people. Note that when a hoist is employed, the lifting-holding-carrying components of the materials handling job are replaced with pushing-pulling components.

3.1.2 Lift Tables

Lift tables function to lift or lower an object and hold it in position so that it can be grasped or manipulated (see Figure 2.6). They can often be used to improve the vertical multiplier (VM) and the horizontal multiplier (HM) for a lifting task. They can also reposition the work area to allow manipulation and assembly operations to be done at an optimal work height. Some lift tables have a rotating and/or tilting top to help workers position the load closer to their body (thereby reducing the HM). Lift tables should be used when lifting is repeatedly done from a fixed location. They are particularly effective when used for stacking and unstacking operations. Note that lift tables eliminate manual lift and hold tasks, but they often require the worker to manually lift the load from the table and carry it to a different location.

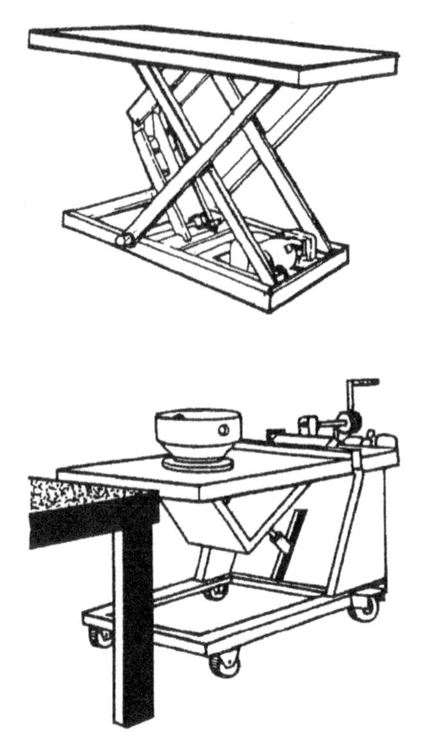

Figure 2.6 Typical lift tables used to lower to raise work heights (from Kroemer et al. 2001)

3.1.3 Conveyors

Conveyors transport a load from one fixed location to another fixed location. They are either gravity powered or powered using an electric motor connected to a belt. Gravity-powered conveyors (see Figure 2.7) use low friction wheels to transport the load along the length of the conveyor. These devices must ensure an even slope from the beginning of the conveyor to the end. As a result, gravity-powered conveyors often result in a less-than-optimal high lift onto the conveyor at the origin and a lower-than-optimal lift off the conveyor at the destination. Electrical conveyor belts do not have these same limitations, but are much more expensive and have pinch and entanglement hazards associated with their operation. Note that conveyors eliminate the carry components of materials handling but usually require the worker to perform the lifting and lowering components of the job. Conveyors are often combined with hoists or lift tables to address both the lifting and carrying components of a work task.

3.1.4 Hand Trucks

Hand trucks or carts (see Figure 2.8) are used to transport loads from a variable location to a different variable location. Unlike a conveyor, they are not tied to a specific beginning and ending location. However, they have the disadvantage of

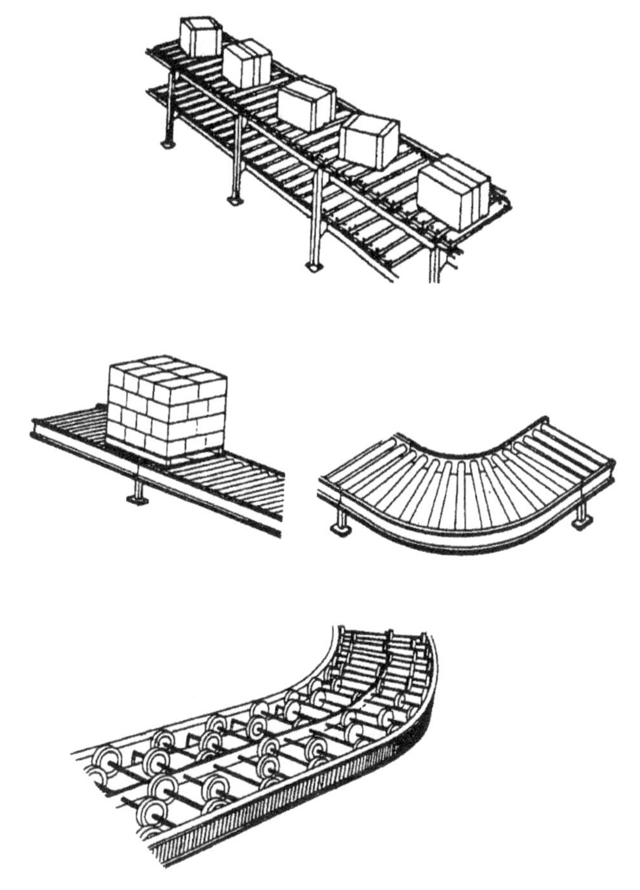

Figure 2.7 Typical gravity-powered conveyors used to move materials from one fixed point to another fixed point (from Kroemer et al. 2001)

requiring the worker to physically push and/or pull the load from one location to another. In some situations, loading a hand truck can be done without requiring the worker to manually lift the load (such as when the lip of a hand truck is slipped underneath a stack of boxes). However, in many other situations, loads will have to be lifted onto the hand truck at or very near floor level, resulting in a hazardous lift (see Section 2.2). Hand trucks replace the carry component of a task with a push-pull component, but usually require the worker to perform the lifting and lowering components of the job. Similar to conveyors, they can be used with hoists or lift tables to address both the lifting and carrying components of a work task.

3.2 Ergonomic Issues in Device Design, Selection, and Installation

Mechanical materials handling devices may solve one ergonomic issue but at the same time create new ergonomics concerns for the worker. As a result, workers

Figure 2.8 Typical hand trucks and carts used to transport materials (from Kroemer et al. 2001)

may choose to ignore mechanical assists or use them improperly in an effort to become more productive.

3.2.1 Time Standards and Cycle Time

The chief complaint of workers whose job is reconfigured to include a materials handling assist device is that the cycle time for the tasks increases significantly. In other words, the time required to attach the load to the device, transport it to the new location, and unattach it from the device is significantly longer than physically picking up the load, carrying it to the new location, and releasing the load. If the device is introduced and time standards are not adjusted accordingly, workers often revert to manual methods in very short order. This can be the case even when time standards are adjusted, because the risk of lifting is not easily judged, while the benefits of additional free time within the work cycle are immediate and apparent. In introducing an assist device, it is imperative that management insist that the device is used on every work cycle and that time standards are appropriately determined to take into account the added time required to use the device.

Although increased cycle time often accompanies the use of an assist device, it has been our experience that novel approaches can many times be used to reduce cycle time. Assist devices should not be viewed as simply replacing human task components, but should be viewed as devices that have unique capabilities and limitations. For example, most hoist devices have load carrying and manipulation capabilities that far exceed the capabilities of a worker. Where a worker may have picked up a single part, transported it to the assembly line, and fixed it to the assembly, a hoist may be capable of loading and transporting many parts at once. Clever end effectors can be designed to load multiple parts, thus reducing the number of transports involved in the job and more than compensating for the extra time to attach and release the parts.

3.2.2 Horizontal Pushing and Pulling

Many materials handling assist devices replace lifting and carrying job components with horizontal pushing and pulling job components. Although these job components are usually less hazardous, as compared to lifting and lowering tasks, they can involve significant force and effort. Forces up to 500 N, or 80 percent of the worker's maximum strength capability, have been recorded while using a materials handling assist device (Resnick and Chaffin 1997; Woldstad and Chaffin 1994). Similar to lifting, pushing and pulling can be hazardous activities depending on the work configuration (see Chaffin et al. 2006 or Mital et al. 1997 for a more detailed treatment of the ergonomics of pushing and pulling).

Among the factors that increase the force required to push and pull a device are the mass of the system and the internal joint friction (Woldstad and Chaffin 1994). In general, people are very poor judges of mass and inertia. As a result, they accelerate very quickly and do not allow themselves enough time and space to decelerate the load and come to a stop. High deceleration forces and a tendency to overshoot the desired stopping location are common with high inertia loads. To address this problem, materials handling devices should be designed and configured to minimize the mass that is being moved or transported.

In addition to mass, the joint friction should be carefully considered within each device. If the joint friction is too high, increased push and pull forces will result. However, if the friction is too low, the inertial mass of the system will be hard to control, as already described. An intermediate value that allows easy control of the system and resists inadvertent movement, while minimizing push and pull force, is desired. This is often the result of careful tuning and adjustment of the device once it is installed.

Push and pull forces will be limited by the coefficient of friction between the floor surface and the workers' shoes. Coefficient of friction is defined as the ratio of the vertical force to the horizontal force acting on the surface. For pushing and pulling, the vertical force is determined by the weight of the person. If the person tries to exert a large horizontal force, the coefficient of friction of the shoe–floor interface must be high enough relative to the weight of the person to support the activity. If the coefficient of friction is not high enough, the person's foot will slip, with the potential for a fall and/or loss of control of the load. Ensuring a high coefficient of friction for all floor surfaces upon which pushing and pulling is done is an important safety consideration for the use on materials handling assist devices.

3.2.3 Device Control

Materials handling assist devices should be designed in a manner such that it is clear to the user of the device how a manipulation of the device controls will result in movement. This is true both when the device is manipulated by hand (though pushing and pulling) and when it is controlled though a motor and a control system. In designing an assist device, there is a tendency to overcome the physical constraints to movement through the addition of more complex segments or links. Particularly for hoists, this can result in systems where there are more degrees of freedom than needed. In such a system, the device can be positioned to a location in space with more than one configuration. Such systems are inherently hard for people to use and should be avoided. For any device, the user should be able to predict with certainty what direction the device will move when any component is physically pushed or pulled.

With respect to automatic controls, care should be taken to consider device control-response compatibility. If an operator is holding a push button controller, the expectation will be that the right button moves the device to the right; the top moves the device upwards, and so on. This is a challenge if the controller is separate from the device, as is often the case for overhead hoist devices. For these devices, the orientation of the worker to the device (front, back, right side, left side) will determine the control–response relationship. What works appropriately on one side of the device will likely work inappropriately on the other side. Designers should try to anticipate where workers will stand when operating the device and make sure actuation is appropriate for that position. Redundant labels for the keys of these devices with coding relative to the room up-down, north-south is often a good idea to help users orient in unusual work positions. In addition to orientation, controls should be designed with moderate system gain or responsiveness. As a rule, human operators have difficulty controlling automated devices that have both a long control delay and a high system gain. Material handling assist devices often have a considerable delay in their control system, making attention to gain an important design consideration.

3.3 Device Design Summary

Material handling assist devices can be an effective way to reduce the risk to workers associated with manual materials handling. When installing such devices, engineers should be cognizant that some devices address the lifting component of the task and others address the carry component. Effective design will often require that both types of devices be employed. When selecting devices, remember that that the best devices have simple kinematics and low movement mass. They will also have control systems that are easy to operate and are compatible with human expectations. Finally, operators are likely to take longer to do a task when using a manual materials handling assist device and appropriate work methods changes to accommodate this are needed for successful adoption.

4 IMPLICATIONS OF USING AN ERGONOMIC APPROACH

The use of the NIOSH guidelines to evaluate your workplace, followed by mechanical assists to improve the work environment, is an example of the power of an ergonomic approach. *Ergonomics*, the science of designing a job by considering the characteristics and limits of people, allows individuals to maximize their productivity and minimize their exposure to harmful stressors, helping to improve both process and product design. Proper use of ergonomics has a positive effect on many business costs and, so, the financial bottom line. But what are the benefits? What are the different options available to the manager?

4.1 Value of Ergonomics

Worker musculoskeletal disorders (MSD) continue to plague manual material handing (MMH) jobs. In 2002, the U.S. Bureau of Labor Statistics reported over 58,000 cases of repetitive motion injuries and, in 2004, over 208,000 cases that involved overexertion injuries because of heavy lifting (Tahmincioglu 2004). Ergonomics can play a significant role in reducing these numbers.

What benefits can be realized by using an ergonomic approach in the workplace? Hendrick (2003) categorized ergonomic benefits into three areas: worker-related benefits (such as reduced accidents and injuries, and reduced skill requirements), productivity benefits (such as increased output per worker and reduced errors), and materials and equipment benefits (such as reduced maintenance, scrap, and equipment damage). Management can easily track all of these benefits to show improvement.

One of the principles of ergonomics is that every ergonomic improvement should also result in a productivity improvement. Conway and Svenson (2001) examined the relationship between MSD and workplace productivity. They concluded that lower MSD rates were significantly correlated with higher productivity increases, and noted that industries having high MSD rates had lower productivity gains. Their study showed that some industries and workplaces are much more likely to be beset by MSD injuries, and in need of innovative ergonomic solutions, than others.

The value of using an ergonomic approach to attack problems in the workplace can be illustrated by the NIOSH guide, an ergonomic tool that is both diagnostic and prescriptive. That is, it may be used to find a problem as well as find a solution to that problem. The guidelines themselves do not tell you how to fix the problem (the solution), but provide you with task characteristics that pinpoint areas for improvement.

The finding of simple, cost-effective solutions (such as found in Section 2.3) is often a stumbling block to workplace improvement.

4.2 Controls and Solutions: Fitting the Job to the Worker

Once MSD risk factors have been identified, and properly measured (say, by the NIOSH equation (5)), an appropriate control strategy must be chosen. The goal of your solution is to match task demands with worker capabilities, improving the fit of the job with the worker. As introduced at the beginning of this chapter, many different types of ergonomic interventions are available for consideration, falling into three broad categories: administrative, engineering, and providing personal protective equipment (PPE).

Engineering solutions involve a change in the physical features of the workplace. Engineering controls redesign the job to minimize effects on the worker and are often permanent. This includes approaches such as automation (this redesign eliminates human exposure), using mechanical aids (as presented in Section 3), and modifying the job through redesign (adjusting the work process or workstation).

When the cost or feasibility of engineering controls become unreasonable, administrative solutions, which focus on the worker, offer methods to reduce the exposure of workers to a hazard. Administrative controls are the workplace policies, procedures, and practices that minimize the exposure of workers to hazard conditions. They are less effective than engineering controls, as they do not usually eliminate the hazard, instead lessening the duration and frequency of exposure to the hazard condition.

If administrative controls are not appropriate, personal protective equipment (PPE) should then be considered. They are the least effective form of control, as they do not eliminate the hazard or reduce the time of exposure. PPE reduces hazardous exposure by placing a barrier between the hazard and the worker, such as through the use of gloves or earplugs.

The most effective method of reducing or eliminating ergonomic hazards is to fix the hazard, not the worker. Focusing on engineering controls for reducing injuries has several advantages over the worker-focused administrative approaches. They do not depend on worker capabilities (such as strength, motor skill, and conditioning) to prevent injury. Temporary or new workers may become injured because they do not have similar physical characteristics to meet the physical demands of the task, as did the original worker.

Although the final solution to any materials handling problem may involve a combination of workplace and worker adjustment, early focus should not be on the worker but on developing workplace solutions.

5 SUMMARY: THE NIOSH LIFTING EQUATION, ERGONOMICS, AND YOUR BOTTOM LINE

The benefits that can be realized by using an ergonomic approach in the workplace include worker-related benefits, productivity benefits, materials and equipment benefits, providing valuable financial incentives to companies. The use of one ergonomic tool, the NIOSH lifting equation, provides a dependable physical evaluation of a task. When coupled with an appropriate solution, a company can realize many of these benefits.

To be successful, employers need to identify, assess, modify, or eliminate handling tasks they require workers to perform. The lifting equation is a useful tool for assessing the physical stress of two-handed manual lifting tasks. The NIOSH guide for lifting, developed by considering epidemiology, biomechanical models, physiological limits, and psychophysical data, is limited to the conditions for which it was designed. The guidelines serve two functions—diagnostic and prescriptive. That is, the results provide a diagnostic peek into the stresses on the worker and, at the same time, provide information on what needs to be improved. Solutions can then be developed based on the analysis. The lifting equation is only one tool in what needs to be a comprehensive effort to prevent work-related low back pain and disability.

Engineering controls are the bread-and-butter of ergonomic solutions. They change the workplace to fit the task to the worker, often resulting in a design that accommodates a wide range of workers. The NIOSH guidelines focus us on engineering solutions, providing a tool that will allow employers to provide a better workplace, reduce MMH injuries, and increase productivity.

REFERENCES

Ayoub, M. M., N. J. Bethea, S. Deivanayagam, S. S. Asfour, G. M. Bakken, and D. Liles. 1978. Determination and Modeling of Lifting Capacity, Final Report, DHHS (NIOSH) Grant No. 5-R01-0H-00545-02.

Chaffin, D. B., G. B. J. Andersson, and B. Martin. 2006. *Occupational biomechanics*, 4th ed. New York: Wiley.

Conway, H., and J. Svenson. 2001. Musculoskeletal disorders and productivity. *Journal of Labor Research*, 22 1: 29–54.

Hendrick, H. 2003. Determining the cost–benefits of ergonomics projects and factors that lead to their success. *Applied Ergonomics* 34 (5): 419–427.

Kroemer, K., H. Kroemer, and K. Kroemer-Eberts. 2001. *Ergonomics: How to design for ease and efficiency*, 2nd ed. Upper Saddle River, NJ: Prentice Hall.

Mital, A., A. S. Nicholson, and M. M. Ayoub. 1997. *A guide to manual materials handling*, 2nd ed. London: Taylor & Francis.

NIOSH. 1981. *Work practices guide for manual lifting*, DHHS Publication 81-112. Washington, DC: Government Printing Office, 20402.

Resnick M. L., and D. B. Chaffin. 1997. An ergonomic evaluation of three classes of material handling device (MHD). *International Journal of Industrial Ergonomics* 19 (3): 217–229.

Smith, J. L., M. M. Ayoub, J. L. Selan, H. K. Kim, and H. C. Chen. 1989. Manual materials handling in unusual postures. *Advances in Industrial Ergonomics and Safety I*. Taylor & Francis.

Smith, James L., M. M. Ayoub, and Joe W. McDaniel. 1992. Manual materials handling capabilities in non-standard postures. *Ergonomics*, 35 (7/8): 807–831.

Snook, Stover H. 1978. The Design of Manual Handling Tasks. *Ergonomics* 21: 963–985.

Snook, Stover H., and V. M. Ciriello. 1991. The design of manual handling tasks: Revised tables of maximum acceptable weights and forces. *Ergonomics* 34: 1197–1213.

Tahmincioglu, Eve. 2004. Ergonomics is back on the radar screen for both business and regulators. *Workforce Management* 83 (7): 59–61.

Tayyari, Fariborz, and James L. Smith. 1997. *Occupational ergonomics: Principles and applications*. Boston: Klewer Academic Publishers.

Waters, Thomas R., Vern Putz-Anderson, Arun Garg, and Lawrence J. Fine. 1993. Revised NIOSH equation for the design of manual lifting tasks. *Ergonomics* 36 (7): 749–776.

Waters, Thomas R., Vern Putz-Anderson, and Arun Garg. 1994. *Applications manual for the revised NIOSH lifting equation*, DHHS (NIOSH) publication No. 94–110, U.S. Department of Commerce, National Technical Information Service (NTIS), Springfield, VA, 22161.

Woldstad, J. C., and D. B. Chaffin. 1994. Dynamic push and pull forces while using a manual material handling assist device. *IIE Transactions* 26 (3): 77–88.

Woldstad, J. C., and R. Reasor. 1996. In *Occupational ergonomics: Theory and applications*, ed. A. Bhattacharya and J.D. McGlothlin. New York: Marcel Dekker.

CHAPTER 3

INTELLIGENT CONTROL OF MATERIAL HANDLING SYSTEMS

Kasper Hallenborg
The Maersk Mc-Kinney Moller Institute
University of Southern Denmark
Odense, Denmark

Manufacturers in highly developed countries around the world have, during the last decade, experienced new challenges due to globalization and changes in customer requirements. Shortening the time to market for new products and user customization are some of the factors that challenge production planning for many companies. Mass production of highly standardized products is either moving to low-wage countries or is being replaced by more sophisticated alternatives required by more demanding consumers.

Mass customization is the new concept for manufacturers, introduced by Stan Davis (1989), that challenges the traditional neoclassical economical model of customers as rational consumers who seeks to maximize their benefits and minimize their costs. The growth in communication technology, globalization, and improved economy of consumers has shifted the decision-making power from the producers and the governments to the customers.

This chapter will start with an introduction to a more decentralized approach for controlling systems for manufacturing and material handling. Different approaches for intelligent control will be discussed, and finally two cases of material handling will be presented—one large-scale complex system of baggage handling in an airport and the other a case of scheduling items through a manufacturing process.

1 HISTORICAL INTRODUCTION

Beginning with the Industrial Age, high-volume, low-variety products were the new trend among manufacturers, resulting in low-cost, high-quality products. To begin with, customers were satisfied by the new opportunities realized by mass production, even though customer requirements were not the driving force in product design. Due to low competition in markets, manufacturers were more concerned with production efficiency than customer requirements (Sipper and Bulfin 1997). Likewise, dominating management theories of that time focused on rationalization, such as Taylor's scientific management (Taylor 1911).

Improvements in automation technologies led manufactures to see the possibilities of exchanging labor-intensive tasks with specialized machines and material handling systems to rationalize production. The automotive industry was among the first to take advantage of automation: Oldsmobile Motor introduced a stationary assembly line in 1907, followed by a moving assembly line in 1913 at Ford's new factory in Highland Park, Michigan, even handling parts variety (Sipper and Bulfin 1997).

For decades, mass production, automation, rationalization, and scientific management were the dominating factors in manufacturing, but that gradually changed towards the end of the twentieth century. Especially due to the growth in international competition, market demands pushed forward new challenges for manufacturing—flexibility and customization. Companies in Japan were the first to address the new conditions; they changed from mass production to lean production. Instead of focusing on having high volume and rationalization as the key drivers in developing mass-production environments, lean production focuses on the whole process of production—eliminating inventory, decreasing costs, increasing flexibility, minimizing defects, and creating high product variety.

As trends in the automotive market changed, customers were no longer satisfied by standard cars, but required customization (Brennan and Norrie 2003). Davis was the first to introduce the concept of mass customization, which tries to combine the low unit cost of mass-produced items with flexibility required to produce customized by individual customers through computerized control of production facilities.

2 FLEXIBLE MANUFACTURING

As flexibility was commonly accepted as one of the primary nonfunctional requirement for new manufacturing systems, research and development initiatives naturally concentrated on means and technologies to cope with the new demands.

The notion of a *flexible manufacturing system* (FMS) was born when Williamson in the 1960s presented his System24, a flexible machine that could operate 24 hours without human intervention (Williamson 1967).

Computerized control and robotics were promising tools of the framework for automation, which could increase flexibility. Obviously, not all products or systems would benefit from or require increased flexibility, but FMS was intended to close the gap between dedicated manufacturing hardware and customization, as outlined by Swamidass (1998) in Figure 3.1.

FMS has the advantages of zero or low switching times, and hence is superior to programmable systems. Despite that, however, FMS has had only limited success in manufacturing setups.

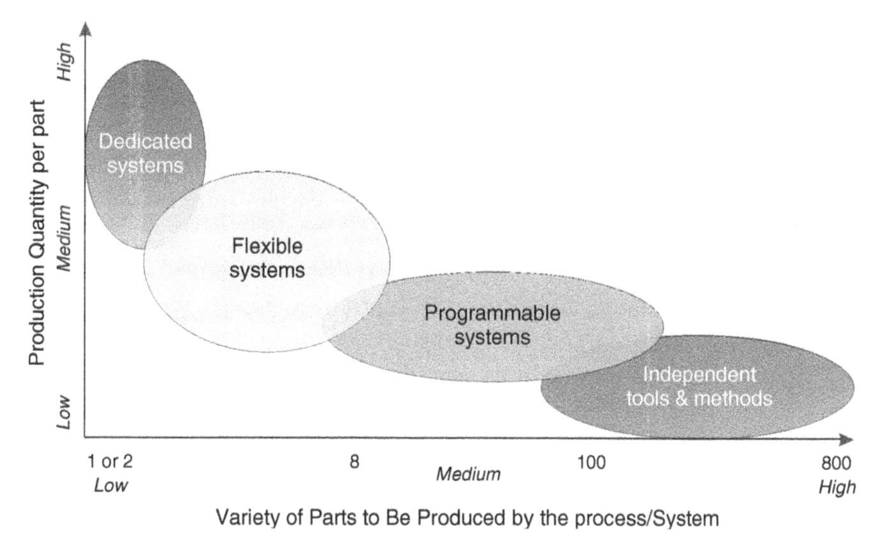

Figure 3.1 Manufacturing flexibility spectrum (adopted from Brennan and Norrie 2003)

Systems integration is the main issue for FMS to be successful, and flexible hardware and manufacturing entities is only one part of the answer. The control software to handle and integrate the flexible entities in the overall process is equally important (Brennan and Norrie 2003).

In fact, the control software is often regarded as the critical part, as it requires high expertise from developers. The complexity of the system and time-consuming process for reconfiguration have often led to low understandability of the system, which is an important problem to manufacturers, who are not experts in manufacturing technologies.

FMSs are often composed of computer-aided or robotic assembly nodes, which are connected by some form of material handling system. Each cell can automatically handle either planned or unpredicted changes in the production flow.

The centralized control generally used in FMSs—which are based on principles and algorithms of classical control theories—would not scale very well for such large systems as identified by Sandell et al. (1978). That was the main issue leading to new approaches for manufacturing control. Bussmann (1998, p. 3) was even more specific and clear in his conclusion:

> Manufacturing systems on the basis of CIM (Computer Integrated Manufacturing) are inflexible, fragile, and difficult to maintain. These deficits are mainly caused by a centralized and hierarchical control that creates a rigid communication hierarchy and an authoritarian top-down flow of commands.

3 DISTRIBUTED SYSTEMS

The experienced problems with complexity and maintenance led to new approaches in the area of manufacturing control. Parunak (1995) states that traditionally a centralizing scheduler is followed by control, which would generate optimal solutions in a static environment, but no real manufacturing system can reach this level of determinism. Even though scheduling of a shop floor environment could be optimized centrally, the system would fail in practice to generate optimal solutions due to the dynamic environment caused by disturbances such as failures, varying processing time, missing materials, or rush orders (Brennan and Norrie 2003).

In general, rescheduling and dissemination of new control commands are time consuming and bring the centralized model to failure. Instead, Parunak (1996) argued that manufacturing systems should be built from decentralized cooperative autonomous entities, which—rather than following predetermined plans—have emergent behavior spawned from agent interactions. He listed three fundamental characteristics for a new generation of systems:

1. Decentralized rather than centralized
2. Emergent rather than planned
3. Concurrent rather than sequential

The area for intelligent manufacturing systems was born, and research was conducted in different directions. One of the major approaches was a project under the intelligent manufacturing systems (IMS) program, called holonic manufacturing systems (Christensen 1994), which settled as a new research area for manufacturing control. Holonic systems are composed of autonomous, interacting, self-determined entities called *holons*.

The notion was much earlier introduced by Koestler (1967) as a truncation of the Greek word *holos*, which mean whole. The suffix on that means part, similar to the notion used for electrons and protons. Thus *holons* of the manufacturing entities are parts of a whole.

The HMS project was initialized by a prestudy (Christensen 1994), before the large-scale project was launched in the period from 1995 to 2000. The huge initiative had more than 30 partners worldwide. Not only did the project focus on applications, but also three of the seven work packages concentrated on developing generic technologies for holonic systems, such as system architecture, generic operation (planning, reconfiguration, communication, etc.), and strategies for resource management. The application-oriented foci were organized in four work packages concerning manufacturing units, fixtures for assembly, material handling (robots, feeders, sensors, etc.), and holomobiles (mobile systems for transportation, maintenance, etc.).

The project was very successful regarding generic structures of the holons aimed at low-level, real-time processing. The specification of the holons was even formally standardized by the International Electrotechnical Commission (IEC) 61499 series of standards.

The holonic parts came to short in systems requiring higher level of reasoning (Brennan and Norrie 2001), thus the term of holonic agents was introduced (Mařik and Pěchouček 2001). Software agents encapsulate the holon and provide higher-level decision logic and reasoning, but also more intelligent mechanisms to cooperate with other holonic agents.

Generally, agent technologies provide a software engineering approach to analyze, develop, and implement intelligent manufacturing control for distributed entities and holons. Whereas the holons were formally specified through the IEC standards, agent-based manufacturing control still lacks from having formal standards, even though various attempts have been taken—YAMS (Yet Another Manufacturing System) by Parunak (1987) or MASCADA (Brückner et al. 1998) among others.

Research on manufacturing and material handling systems has gradually moved from a monolithic control toward a decentralized, distributed, and—most recently—agent-based approach, but only a few real systems have adopted the shop-floor models. Production 2000+ (P2000+) is an exception, and is generally known as the first agent-based manufacturing system that has moved from research into real production. P2000+ was installed at Daimler to produce cylinders. The objectives of the project were to develop a robust and flexible

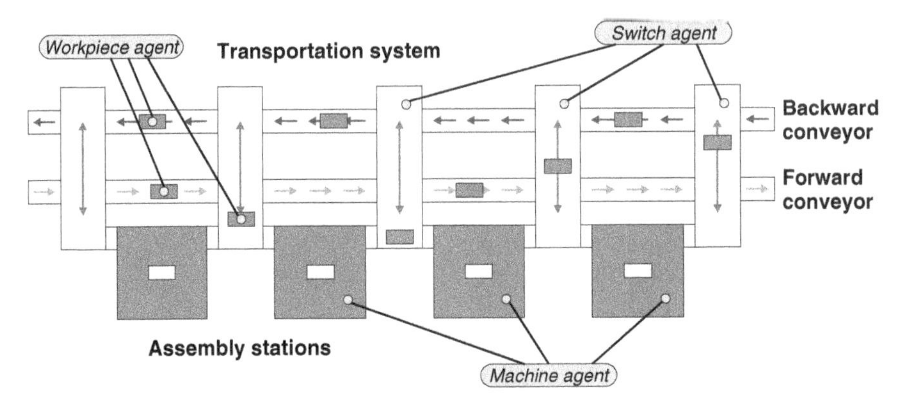

Figure 3.2 Illustration of the system layout and agent mapping in Production 2000+, adopted from Bussmann 2001

manufacturing system, through a set of flexible machines that were connected by a flexible material handling system (Bussmann and Schild 2001). An overview of the P2000+ system layout and agent mapping is illustrated in Figure 3.2.

Before going into details about agent technologies and how they can be applied to manufacturing and material handling systems, this chapter will address the issues of the introduction. The new conditions for manufacturers that push toward more intelligent control will lead to new objectives for the design of such systems, and it is important to know if they fit with an agent-based approach.

4 NEW CHALLENGES

Volatile markets, globalization, emergent technologies, and increasing customer requirements are pushing new challenges to manufacturers. Shen and Norrie summarize a number of fundamental requirements that must be considered for the next generation of manufacturing systems (Shen and Norrie 1999):

- *Enterprise integration*. With constantly changing market and user requirements, the time to market is decreasing. Thus, a competitive manufacturing system must be integrated with related management systems, so purchasing, orders, personnel, materials, transport, and so on are taken into account.
- *Interoperability*. The information environment for new systems can no longer be expected to be homogenous and from the same vendor. Systems may be composed of subsystems, which must cooperate and interact.
- *Open and dynamic structure*. New subsystems could even be added during operation, which require open and dynamic architectures that allow new components to be integrated regarding both software and hardware.
- *Cooperation*. Cooperation must be established with customer, suppliers, and other partners in order to secure the flow of materials and discover all customer requirements.

- *Agility*. Agile manufacturing is a key concept. It is the ability to quickly adapt and reconfigure the manufacturing environment to unanticipated changes.
- *Scalability*. It is important that the organization can grow or shrink at any level when required.
- *Fault tolerance*. Environments are not static. Failures will occur once in a while; therefore the system must be able to detect and recover from such system failures.

The worldwide trend toward low-batch production with an increased variety has been ongoing for several decades. The growth and advancements in communication technologies have enabled customers to raise the individual desires or even take part in the development process. Information technology has also made it much simpler to get valuable customer feedback, which can evolve and improve the products—thereby also shorten the product lifecycle.

Previously manufacturing sites were optimized by a linear production model that was suitable for mass series production, such as transfer lines, but long switching times make them inherently less suitable for flexible manufacturing, resulting in low utilization.

Both flexible and distributed systems, as presented in the previous section, can meet many of the requirements mentioned, but fail on others as well. Holonic systems focus on creating flexible systems through decentralized and cooperative components, which will benefit the systems with respect to agility and scalability of fault tolerance, but the heterogeneous environment and enterprise integration are given no special attention. Distributed systems might solve the issues of interoperability, as well as open and dynamic structures, but at the local scope there is no guarantee for flexibility and efficiency.

Computer integrated manufacturing (CIM) has also been proposed as a solution to cope with the new challenges. CIM is an approach where the entire production process is controlled by a computer. It is organized in a hierarchical architecture from the strategic level of the company to the production level, but with closed-loop control so that feedback is provided back to the subsystems in order to optimize the entire process (Asai et al.1994, Rembold and Nnaji 1993). However the centralized and sequential approach to control both planning and scheduling was found insufficiently flexible and agile for the dynamic production environment and the changing production styles (Colombo 2006). Huge investments were required to install the sensory feedback at the physical machine level and implement them in the centralized monolithic control system. The complexity of the system made them rigid and inflexible, and often conditions had changed when the systems were fully operational. The organization of CIM factories is commonly hierarchical, so a single point of failure could shut down the entire system. For these reasons, the original approach of a CIM factory was never successful in real life (Scheer 1992), as the new requirement of flexibility, dynamic production environment, and mass customization overtook the efficiency of CIM. Instead, the responsibility in CIM systems was distributed to autonomous, intelligent, and collaborating components, which led to the holonic manufacturing approach under the IMS program already mentioned.

The shortcomings of the holonic manufacturing gradually prepared the agent-based approach as the most promising software technology for intelligent control. Whereas holonic systems have the focus on all the mechatronic components of an IMS, an agent model of the system also incorporates the planning, scheduling, and interoperability among the agents.

Multiagent-based systems (MAS) are still a relatively new paradigm in computer science, which can suit many other purposes than control of logistics and manufacturing systems. MAS facilitate an optimization of the decision process and add an extra level of robustness and stability to complex, heterogeneous systems. Agents are able to interact in dynamic, open, and unpredictable environments with many actors—here called *agents*, who cooperate to solve specific tasks or achieve design goals.

Usually, an agent-based manufacturing and material handling system is modeled with agents in all the decision points of the systems, such as assembly stations, employees, cranes, AGVs, robotic cells, PLCs, and so on. These agents will be able to act autonomously by observing their own local neighborhood and communicate with other agents. An agent will also act and change its actions according to the current status of its environment, so it can achieve its design goals as best as possible. This makes the system robust to local unpredictable events, as they will only be perceived by the relevant agents, who will adjust their actions, and the effect will propagate throughout the system through the interagent communication.

5 AGENT TECHNOLOGY

Agent-based, or multiagent, systems had emerged from artificial intelligence long before they were considered for control in manufacturing processes. The research area of artificial intelligence was born in the late 1950s and was focused on both understanding the human reasoning process and developing methods and tools to built intelligent systems. In the first decade, expert systems were the primary base for research in artificial intelligent systems. The decision process of the systems was usually modeled as condition-action rules that were triggered by events from the environment or changes in an internal world model.

Pattern matching and understanding natural languages were hot topics of the time for such type of systems. It was natural to have different knowledge sources that work on different aspect of the problem, which again led to the notion of distributed artificial intelligence (DAI). Different expert systems were used to partially process recorded data in the HEARSAY speech understanding system (Erman and Lesser 1975). Erman and Lesser used a blackboard architecture to combine partial results to find the overall solution to the problem. The different knowledge sources each represent a different aspect or hypothesis on the problem. These are connected to the blackboard and can modify and update the current solution through the shared memory, which contains the definition of the problem.

In the following years, research led to new approaches for distributed problem solving. The contract-net protocol introduced by Smith (1979) was a turning point in DAI. In contrast to the blackboard model, the contract-net protocol has a managing node, which through messages propose a task to the different knowledge sources, which each bid on the task. The manager decides which (could be several) of the contracting nodes can carry out the task and eventually return the result to the manager.

Hewitt was interested in the modeling aspects of distributed problem solving and introduced the actor model (Hewitt 1977). Actors are computational entities with both a script that defines the actions and a list of other actors it can contact—the so-called acquaintances. In the model, actors are awakened when they receive messages from others actors. The actor then runs its script, will die, and is subsequently removed by the garbage collector. During execution of the script, it can both spawn new actors and send messages to its acquaintances.

Given the concepts of message passing and well-defined behavior through the action script, the actor model was a natural predecessor to the multiagent paradigm.

Multiagent systems are appropriate for studying and managing dynamical and heterogeneous systems. They are an approach to handling the increasing complexity of centralized systems by breaking them into simpler tasks, which also give a more natural modeling approach.

Starting with the simplest nonintelligent agents, there exist no commonly agreed definition of an agent, but it is generally accepted that it involves some kind of autonomy, which means that the agent is allowed to choose its own action. Also, the notion of being in an environment is central to agents, as they base their actions on sensory impressions from the environment, which could be either physical or virtual environments. Thus, some sort of input function or perception unit is required for an agent. One of the most cited definitions of an agent is given by Wooldridge:

> An agent is a computer that is situated in some environment, and that is capable of autonomous action in this environment in order to meet its design objectives (Wooldridge 2002).

An agent can be as simple as your heating thermostats, but intelligent agents are the most interesting to be studied in both agent and multiagent systems. Figure 3.3 presents the classic illustration of an agent situated in its environment.

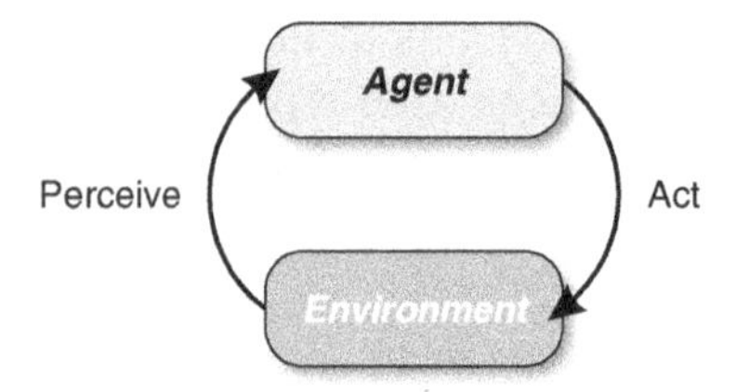

Figure 3.3 Classic agent illustration

There might not be some clear distinction between intelligent or nonintelligent agents, but intelligence is usually combined with some sort of learning mechanism, and Wooldridge and Jennings add these properties to the classification of an intelligent agent (Wooldridge and Jennings 1995):

- *Social*. Intelligent agents can interact with other agents and systems, including humans.
- *Proactive*. Agents not only respond to their environment, but they can exhibit goal-directed behavior on their own initiative.
- *Reactive*. Agents are capable of perceiving changes in their environment, and react upon that.

Other characteristics, such as adaptability, mobility, and rationality, just to mention a few, have been related with the agent model, but is it commonly accepted the four concepts just described (autonomous, social, proactive, and reactive) are the main characteristics of an agent.

6 MULTIAGENT SYSTEMS

Single-agent systems are very close to traditional centralized monolithic systems, but the real strength of agents is revealed when multiple agents interact. Therefore a multiagent system (MAS) is defined as follows:

> A multiagent system is a collection of interacting agents (Bussmann et al. 2004).

It is important to state that a MAS is far more than just a collection of agents, as it also encompasses the emergent behavior, which spawns from the organization of agents and their interaction under the influence of the environment in which they are situated. Silva and Demazeau (2002) proposed the *Vowels* formalism to model a multiagent system:

MAS = A (Agents) + E (Environment) + I (Interactions) + O (Organization)

- *Agents*. Agents are the classic entities to consider, when developing a MAS system, as agents are determined to be the local actors carrying out the tasks of the system. Agents are the key concept that comprises the system, which follows the characteristics of single agents.

- *Environment*. The environment is the space in which the agents exists, moves, and interacts. The space could be virtual, informational and conceptual, but typical the environment is represented by a model of the physical space of the MAS community.
- *Interactions*. Interactions and communication are evident in MAS, due to the aspects of distribution in the systems, and originally by Wooldridge and Jennings as the social ability of an agent (Wooldridge and Jennings 1995). Interactions in the MAS community could take many forms; negotiation, collaboration, coordination, queries, or generally any kind of information exchange between agents. Interactions could be formed as abstract speech-act messages, or in other models as simple natural forces that influence other agents.
- *Organization*. Similar to humans, agents can benefit from being organized, either explicitly defined in classic organizational structures, or the organization could emerge from simple interactions among the agents. Organization often serves the purpose of grouping agents with similar or related actions or behaviors. Organizations can be helpful to support agents in planning, performing actions, requesting information, or realize global goals of the agent system.

In the research of multiagent systems, two different perspectives are dominating—either a micro-level or macro-level perspective. For example, the system could focus on the micro-level issues, such as the internal of the individual agents. What is the decision logic of the agent, how will the agent learn, and how can it be ensured that the agent will act autonomously? In the macro-level perspective the multiagent research community is more concerned with the organization of agent, and how the agent will interact and collaborate in an efficient way. Whereas the micro-level perspective is commonly inspired by biological systems, such as the human brain and ant colonies (Parunak 1997), the macro-level perspective analogies are coming from human organizations and societies (Ferber et al. 2003, Zambonelli et al. 2000).

7 AGENT TYPES

Agents are usually classified as being either *reactive* or *deliberate* in their behavior (Bussmann et al. 2004). Coming out of the artificial intelligence community, a deliberate or cognitive behavior of agents was expected:

- *Cognitive or deliberative agents*. A cognitive agent is one that owns a knowledge base and holds a model of the current environment as it has perceived it, but it will not act only on a search in the knowledge base. It has planning capabilities, so it proactively can adjust its actions according to its goals, even though the perceived input is not covered by the knowledge base.

For the reasoning capabilities, a cognitive agent needs a representation of itself, the part of the environment in which it operate and exists, but also the agents it has to communicate with. Thus, the internal state given by this world model will very likely influence the decision and current actions of the agent. The BDI architecture by Rao and Georgeff (1992) is the most classic example of a cognitive agent model, based on desires, beliefs, and intentions, which will be described next.

For cognitive agents, their actions should not be seen as direct action of the changes they perceive from the environment, but more as a result of their reasoning on understanding the world in which they are situated.

- *Reactive agents*. For reactive agents, there is expected to be a matching rule of action in the knowledge base for each of the inputs it perceives that will lead to actions of the agent. Actions are a direct reaction on the inputs of an agent. A reactive agent usually has no internal world model, as its actions are fully described by rules or functions of the input. The knowledge base for reactive agents could be a rule base known from expert systems, where conditional rules map to a specific output, or physical or biological inspired functions, such as physical forces—which again impact the environment or other agents. Therefore, reactive agents are less social, and have only little or no direct communication with other agents. Their communication is more indirect through the environment, such as the concept of stigmercy in swarm intelligence (Valckenaers et al. 2002).

8 AGENT ARCHITECTURES

The agent architecture is closely related to the agent type, as the architecture in the internal organization of the agent, which describe how it reasons and reacts to perceived input. The architecture presents a design model of the agent, where the flow of information from input to actions are explicitly defined through basic concepts of the agent, such as perception, goals, and desires.

A number of different architectures have been proposed for agents, and they are often associated with the type of agents that participate in the systems. One of the best examples of an architecture that support the reactive behavior of the agent is the subsumption architecture by Brooks (1986). The principle of the architecture is that an agent has a set of accomplishing behaviors, arranged in a subsumption hierarchy. Each of the behavior maps a given set of input values directly to an output value that affect the actuators of the agent. The behaviors in the hierarchy is arranged, so lower layers represent low-level behaviors and are prioritized over higher layers that represent more abstract behaviors. The subsumption architecture has been found useful for controlling robots and other AGVs, where lower layers represent the basic high prioritized tasks, such as obstacle avoidance, while the upper layers focus on the general goals of the robots, such as going from A to B or exploring an environment. The famous boid model of Craig Reynolds

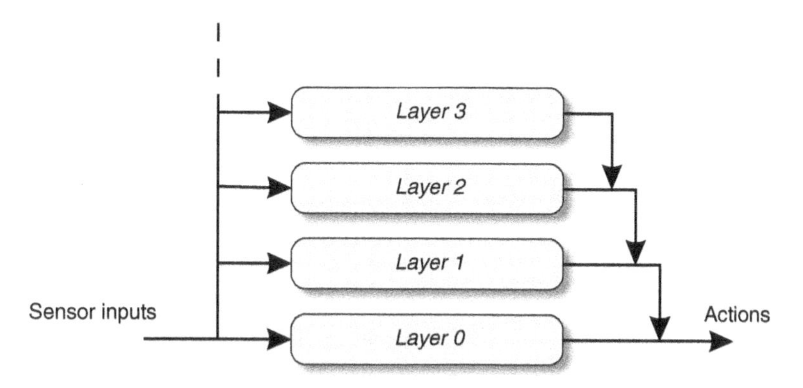

Figure 3.4 Brook's subsumption architecture

to simulation flocking behavior of birds and fish use a similar approach to economize with the energy consumption of the boids (Reynolds 1987). Figure 3.4 shows a model of the layered subsumption architecture by Brooks.

For cognitive and deliberative agents, the most well-known architecture that comprises most the concepts and ideas of highly reasoning and cognitive agents is the BDI architecture (Rao and Georgeff 1992). Here, B is for beliefs, D for desires, and I for intensions, which very well reflect the principles of human reasoning and other intelligent creatures. The agent has a current view of the world or the environment in which it is situated, which is modeled through its beliefs. The goals it has been designed to achieve are described by the desires of the agent. One can think of the desires as a plan library or a set of described goals that the agent wants to achieve, but not necessarily is working on at the moment. So the decision making of the agent works by selecting the desires that seem most achievable under the current conditions (the beliefs). When an agent commits to pursuing a certain desire, the desire becomes an intension of the agent, and the agent persists in pursuing this goal until it no longer appears achievable. Thus, the agent will not just give up on a current plan, whenever new inputs are perceived, so that challenge of using the BDI architecture is to balance the proactive and goal-directed behavior against the influence from new inputs, which is a more reactive behavior. A model of the BDI architecture is presented in Figure 3.5.

Most other architectures for cognitive agents are either an extension of the BDI architecture or they use a somewhat similar approach with a formalized reasoning model between a set of described goals and the current world model.

Hybrid architectures, which try to combine the best of both worlds, also exist, and the InteRRaP architecture (Figure 3.6) by Müller is a classical example of that (Müller 1996). The InteRRaP architecture has three layers:

1. A bottom layer has all the reactive behaviors and situation-to-action rules. This layer is also the interface to the world.

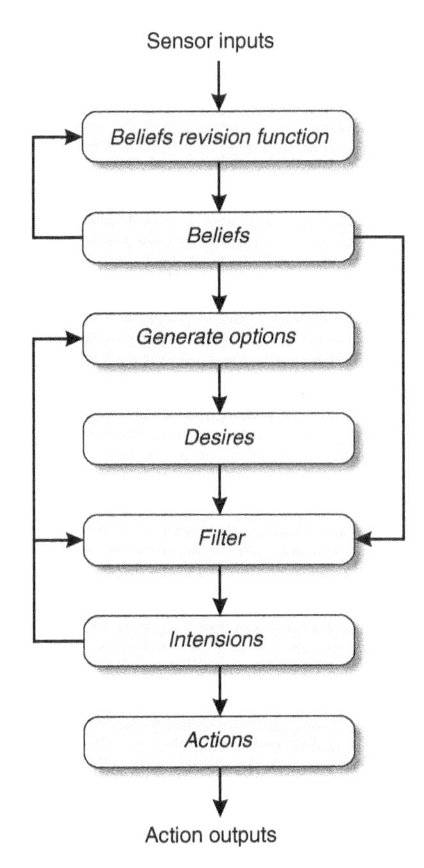

Figure 3.5 BDI architecture

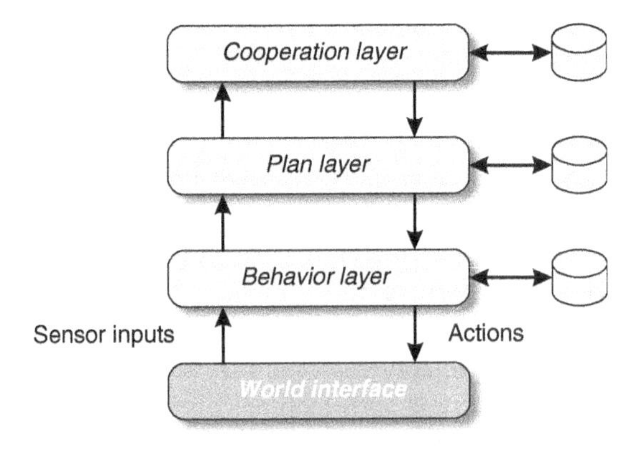

Figure 3.6 The InteRRaP architecture

2. The plan layer is on top of the behavior layer. It handles all the goal-directed and proactive planning of the agent.
3. A cooperation layer describes the collaboration and interactions with other agents.

A common problem with hybrid architectures such as InteRRaP is that there is no clear semantics or methodology for programming such architectures (Wooldridge 1999), so it can be hard for the developer to design a coherent agent behavior.

9 AGENT COMMUNICATION

Interagent communication is a key requirement for an agent to fulfill its social characteristics, so a long list of interaction schemes and communication protocols have been proposed and implemented in the agent community. However, with respect to manufacturing and material handling systems, particular coordination, negotiation, and hybrid mechanisms are dominating (Bussmann et al. 2004).

Apparently, coordination is a form of communication—not only for agents—that will be most noticeable if it does not work properly. Perfectly aligned conveyors, input, and output facilities of material handling systems require a high degree of coordination between the control elements, which are obtained through an often long and tedious alignment process at installation time. For flexible manufacturing systems, the conditions for coordination are constantly subject to changes. Therefore, coordinating agent activities is of highest priority in intelligent manufacturing and material handling systems. With a system composed of flexible cells and connected through, for example, conveyors, coordination can be defined as process of managing dependencies between activities (Malone and Crowston 1994). Coordination is not a trivial thing to achieve in agent-based control systems. Jennings has emphasized three common characteristic of agent systems that lead to dependencies and complicate coordination (Jennings 1996):

- *Actions of agents might interfere*. Two agents might fight for the same resource in order to complete their tasks.
- *Global constraints might have to be satisfied*. It could be suitable to distribute the load on the entire system, but there will still be an overall deadline of an order that an item must satisfy.
- *Individual agents cannot satisfy their own or system goals by itself*. The core idea of a flexible production cell is that an item has to process several work stations for it to be completed.

Malone and Crawston (1994) further simplified the interdependencies relevant for distributed agents into three types of dependencies, as shown in Figure 3.7. A *flow dependency* arises if one task in the process produces or generates a resource that is required by another task. This is the most common dependency

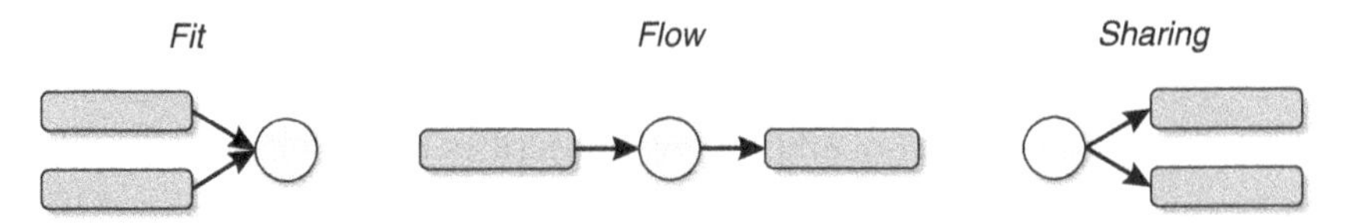

Figure 3.7 Dependency types between task and resources, adopted from Malone and Crawston 1994

in material handling systems (i.e., one conveyor transfers an item to the next conveyor). *Sharing dependencies* occur where several tasks wants to access the same resource. It could be two conveyors diverting from a single conveyor line, where the diverter is the sharing resource that is part of both tasks. A *fit dependency* exists when two or more tasks collectively produce a single resource. The obvious example is, of course, the composition of a complex item, but in a material handling system, fit dependencies also exist when two conveyors line merge to a single line.

Modeling of dependencies and task specifications has been formalized by Decker in the Task Analysis, Environmental Modeling, and Simulation (TAEMS) model (Decker 1996). TAEMS is a framework to model complex computational task environments, either for formalized agent systems or experimental examples. Agents are represented in TAEMS as the executing, communicating, and information-gathering entities, and tree diagram can be drawn to visualize the dependencies and associations among agents.

Decker also proposed generalized partial global planning (GPGP) as a set of generic coordination mechanisms that can bring a decomposed distributed task model of a complex and dynamic environment (e.g., modeled by TAEMS) into a global plan for the collaborating agents. GPGP has been applied to a number of problems related to scheduling, planning, and resource optimization (Decker 1995; Decker and Li 1998; Decker and Li 2000). The GPGP approach has its strengths in reactive planning for agents, which are situated in a dynamic environment. The system should quickly be able to respond to such changes. Each agent optimizes its local plan and synchronizes it with the global plan using a set of standardized coordination mechanisms.

A number of other research papers deal with other approaches for coordination in a complex task environment for agent-based manufacturing (Maionea and Naso 1996; Giret and Botti 2005). Primarily, the research has focused on scheduling flexibility and resource planning in the productions environment under the constraints of the processing steps the different orders have to go through.

Negotiation is the other category of communication principles that are usually applied in agent-based systems for manufacturing and material handling. Negotiation is a well-known concept in sociology for human interaction. Negotiation has motivated many approaches and interaction mechanisms for agents as well. In general, it is about a mutual agreement between two or more actors on some sort of conflicting intensions. Such conflicting interests are what really make humans unique and intelligent, compared to rationally designed systems. Normally, such conflicts can easily be resolved and we learn from it. Pruitt (1981) has provided a clear and also general definition of negotiation:

> Negotiation is a process by which a joint decision is made by two or more parties. The parties first verbalize contradictory demands and then move towards agreement by a process of concession or search for new alternatives.

Naturally, with this relation to social behavior in human societies, many of the negotiation mechanisms have been formalized, adopted, and extended for use in multiagent societies (Bussmann et al. 2004).

The contract-net protocol mentioned in the introduction was one of the first examples of such a mechanism (Smith 1979), where the agent (the manager) that wants a task to be executed proposes it to several other agents capable of performing the task (the contractors). The agreement will be a joint decision between the manager and the contractor, with the lowest offer to carry out the task.

Many of the negation principles are inspired or based on microeconomic principles that provide a formal specification and rational approach to reach a joint decision. Different auction principles are common in negation in agent systems, usually modeled as one agent having a task for one or more agents that calculate a bid to carry out the task. Common auctions methods are single-side, two-side, continuous, and English auctions, where the real design challenge is to find the right interests and goals of an agent to give a fair bid.

Communication principles—such as queries, requests, and publisher-subscriber relationships—are also commonly seen in agent systems, and MAS research has proposed a number of formalized approaches to specify the content of interactions. In general, all agent communication is regarded as message-based interaction, and message content is often presented in an abstract form according to speech-act theories.

For several years, Knowledge Query and Manipulation Language (KQML)was the preferred communication language supported by many agent platforms. KQML was initially specified by the DARPA Knowledge Sharing Effort, led by Finin, and KQML was launched as an interface to knowledge-based systems (Finin 1993).

Later, when the standardization organization for intelligent agents, FIPA (Foundation for Intelligent Physical Agents), announced a number of formal specification for the agent community, ranging from specifications for interaction mechanisms to agent platform architectures, it also specified a new communication language based on the speech-act theory of Searle (Searle 1969). The language is commonly known as FIPA-ACL, which—beside the usual message information, such as the receiver and sender —also allows the message to contain information about the ontology used for encoding the content of the

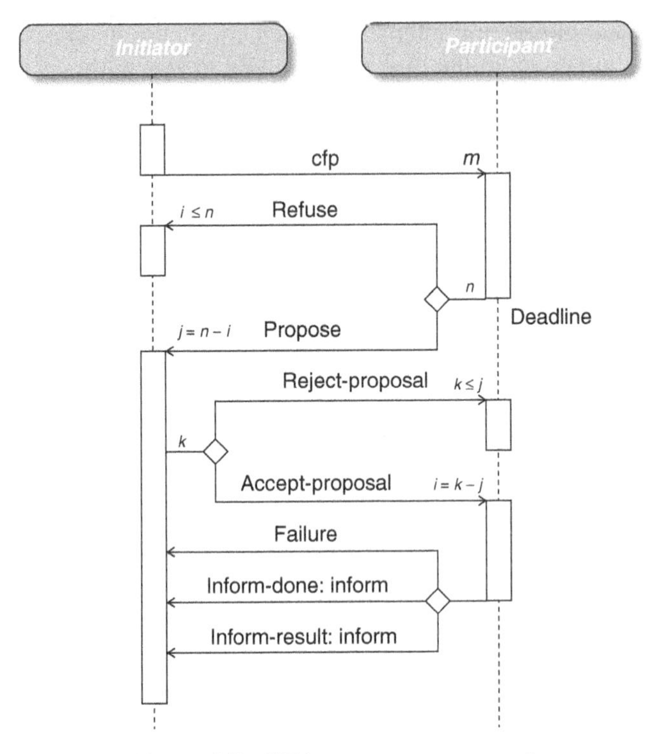

Figure 3.8 FIPA contract-net protocol

message, a time-out indicating the period the sender will wait for a reply, and a performative for the message, which indicates which communication act the message follows. The content field of the message is specified in a semantic language, FIPA-SL. An example of a FIPA specification for the contract-net protocol is given in Figure 3.8.

10 AGENT ORGANIZATION

Whereas the architecture is an internal organization of the components and structure of the agent, organization in MAS will normally refer to a model of the structure and associations among the agents. Horling and Lesser (2005) studied organizational models for multi-agent systems in an extensive survey, where they state:

> The organization of a multiagent system is the collection of roles, relationships, and authority structures which govern its behavior.

They also conclude that no single organizational model will suit all multiagent systems, so they present a list of organizational styles, which are briefly described in the following and illustrated in Figure 3.9.

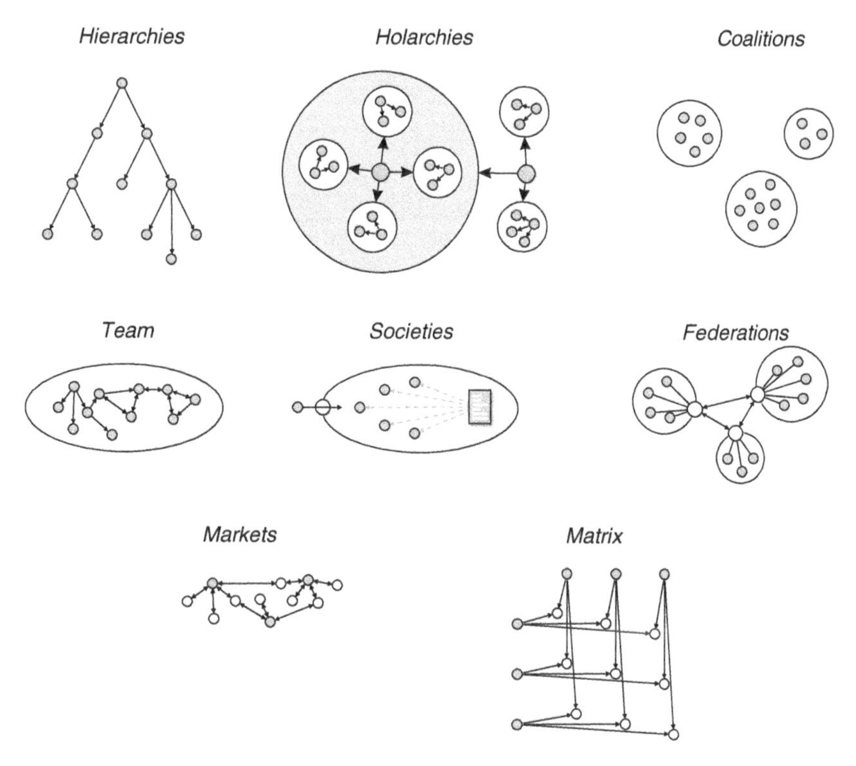

Figure 3.9 Organizational styles (adapted from Horling and Lesser 2005)

10.1 Hierarchies

In a hierarchical model, the agents are arranged in a treelike structure, where direction of action comes to the leaf agents from higher-level agents that have a broader view of the system. Leaf agents collect and provide information for higher-level agents, and horizontal communication is usually not allowed. The strength of the hierarchical style is that parallelism is easy to achieve, and the communication flow is rather limited. The centralized characteristics of the decisional nodes also make it vulnerable for single-point-of-failures.

10.2 Holarchies

Holarchies are based on the structural concept of encapsulation known from the object-oriented paradigm, so almost any entity can be regarded as part of something bigger that can act as a whole. For example, a wheel is a whole by itself, but can also just be regarded as part of a car, and when we deal with the car, it will indirectly affect the wheel. Holarchies can be appropriate to model structural decompositions with autonomy. For manufacturing and material handling it could be different areas of system, which have different responsibility for the process. Obviously, it decreases the predictable behavior of a holon from the outside.

10.3 Coalitions

Coalitions are another structural form known from military and business organizations, where a number of entities join forces on at least a temporary basis. Usually, there will be no internal structure of a coalition, just a flat hierarchy of entities that coordinate their activities to pursue a commonly agreed goal. Coalitions dissolve automatically when they are no longer needed, or a critical number of entities leave the coalition. For material handling coalitions could be use during peak times and/or in case of breakdowns, where parts of the system join forces to cope with the current situation.

10.4 Teams

Teams are also a group of entities that cooperate and coordinate their activities to achieve a common goal, but on a more long-term or even permanent basis. Among the team members there will be some clear representation of the common goals and mutual beliefs that are fundamental for their joint work. The internal organization of a team would typically be a flat hierarchy. On the one hand, a team has the capability to handle larger jobs than a single entity, but on the other hand, communication will increase due to the internal communication inside the team. Teams would also be a typical construct in a material handling system, where parts of the system could work together in bringing an item from A to B.

10.5 Societies

In a society the behavior of agents is governed through a set of social laws, norms, and conventions. Societies are a long-term construct, which group a number of agents that can have quite different goals and heterogeneous capabilities, and the communication will usually also be more diverse and complex. The agents might require extra social skills, and are typically more deliberate than simple coordinating agents. At least from the outside, it is a nice feature that the set of rules and norms are more formalized, so the behavior of agents is more predictable. Societies are not a style commonly used in control of material handling. It would be more suitable to secure flexible interoperability on the enterprise level.

10.6 Federations

A group of agents could form a federation by selecting one group member to represent the group. All communication will go through the delegated agent that will act as a gateway to the outside. It is the responsibility of the delegate to represent and know the individual interests and capabilities of all members, and incorporate them into communication with outside agents. The delegate is also commonly referred to as a broker, facilitator, or mediator. Again, federation is a common style to handle subsystem interoperability by adding an extra agent with the delegate role. The natural disadvantage is, of course, that the delegate will become a candidate for bottlenecks and a single point of failure.

10.7 Markets

Markets are based on the producer versus consumer or buyer versus seller agents principle, where one group of agents (buying or consumers) places bids on shared resources, tasks, or services (the producers or sellers) and the best incoming offer will be chosen. In a market, agents are designed to be competitive, with the potential risk of malicious behavior among the agents, but fairness could also be increased by repeated bidding. The market style is very common for manufacturing systems, as it is rather easy to set up a price calculation function that contains all the relevant factors to prioritize a task, such a deadline, processing time, competences of staff, and similarity with previous items.

10.8 Matrix

A matrix organization is also a common construct in human organizations, where the *worker* agents might have several relationships to different groups or managers. The style is appropriate for project organizations, where the workers belong to different functional groups but at least part time participate in projects led by other managers. The disadvantage is, of course, the potential risk of conflicts, where a worker agent has more managers, but the advantage is that capabilities of the agents can be shared and benefit several places.

11 CASE STUDY 1: BAGGAGE HANDLING SYSTEM (BHS)

It might be clear from the previous sections that multiagent systems span different dimensions on how to classify the system, and in most cases the systems are hybrids, where some part of the system might contain highly cognitive agents that communication a lot as part of their reasoning process, while other parts of the system use simple reactive agents that might solve more trivial tasks.

The two real-life examples that will be presented in this chapter well represent (for real applications) extremes in the space of these dimensions. The baggage handling system, which will be described first, is a complex system of many collaborating and negotiating agents with a cognitive behavior, where the actions of the individual agent are highly dependent on the results of communication with other agents in the systems.

Handling of baggage in airports is shadowed by matters of complexity and uncertainty from the perspective of most passengers, similar to all other issues related to air traffic. Many passengers, frequent or not, feel the moment of uncertainty when watching their bags disappear at check-in counters. Will they ever see their bags again?

Only few imagine the complex system that handles the bags in major airport hubs. Small airports or charter destinations do not fall into this category, but airports with many connecting flights experience this huge sorting and distribution problem. Baggage from check-in is usually not the biggest problem, as the sorting can, to some extent, be handled by distributing flights correctly at the check-in counters. However, bags from arriving planes that have not met their final destination will arrive totally unsorted. So the core task of a baggage handling system (BHS) is to bring each piece of baggage from the input facility to its departure gate. The identity, and hence the destination, of the bags is unknown by the system until scanned at the input facility. This makes the routing principle more attractive than scheduling and offline planning.

A BHS is a huge mechanical system, usually composed of conveyorlike modules capable of transferring totes (plastic barrels) carrying one bag each. The investigated BHS has more than 5,000 modules each, with a length of 2 to 9 meters running at speeds of 2 to 7 meters per second. The conveyor lanes of the modules that make up the BHS in the airport of Munich, Germany, are up to 40 km in total length, and the system can handle 25,000 bags per hour, so the airport can serve its more than 25 million passengers yearly, and the BHS in Munich covers an area of up to 51.000 square meters. Thus, the BHS of Munich is slightly larger than the BHS presented in this case as it has 13,000 modules and more than 80 different types of modules are used, but in setup and control, they are very alike. Later, the different types of modules will be explained when describing how agents have been mapped to the BHS. Figure 3.10 shows a snapshot into a BHS, where a tote containing a bag runs on the conveyors in the foreground.

A BHS often covers an area similar to the basements of the terminals in an airport, and tunnels with pathways connect the terminals. The system is rather vulnerable around the tunnels, because typically there are no alternative routes and the tunnels contain only one or two FIFO-based lanes, which could be several kilometers long. Therefore, the topology of the BHS could look like connected clusters of smaller networks, but within a terminal, the network of conveyors is far from being homogeneous. Special areas, to some degree, serve special purposes.

Besides the physical characteristics of the BHS a numbers of external factors influence the performance:

- Arriving baggage are not sorted, but arrives mixed from different flights and with different destinations, as baggage for baggage-claim are usually separated and handled by other systems.

Figure 3.10 Snapshot into a BHS with a moving tote in the foreground

- Identity and destination of bags are unknown to the system until the bag is scanned at the input facilities; thus, preplanning and traditional scheduling is not an option.
- Obviously, the airport would try to distribute the load of not only baggage, but all air-traffic-related issues over the entire airport. However, changes in flight schedules happen all the time, due to both weather conditions and delayed flights.
- Most airports have a number of peak times during the day, and flight schedules may also differ on a weekly basis or the season of the year. Peak times may influence the strategy on routing empty totes back to the inputs, as they share the pathways of the full totes.

11.1 Performance Criteria

A top priority of a BHS is that no bags are delayed, which can postpone flights and result in airports being charged by airline companies. Therefore, the BHS must comply with the maximum allowed transfer times, in this case between 8 to 11 minutes, depending on the number of terminals to cross. Keeping the transfer time low is a competitive factor among airports, as airline companies want to offer their customs short connections.

Besides ensuring that bags reach their destination in time, the capacity of the BHS should also be maximized, and the control system should try to distributed the load and utilize the entire system if it should be capable of handling peak times. Robustness and reliability are also of top priority, as breakdowns and deadlock situations inevitably lead to delayed baggage, and, in the worst case, stop the airport for several hours.

To fully understand the importance of delayed bags, the concept of *rush bags* must be introduced. Dischargers are temporarily allocated to flights, which define a window where bags can be dropped for a given flight. Normally, the allocation starts 3 hours before departure time and closes 20 minutes before departure. Bags arriving later than 20 minutes before departure will miss the characteristic small wagon trains of bags seen in the airport area. Thus, the system must detect if the bag will be late and redirect it to a special discharger, where these rush bags are handled individually and transported directly to the plane by airport officers. Obviously, this number should be minimized, due to the high cost of manual handling.

Bags entering the system more than 3 hours before departure are not allowed to move around in the system waiting for a discharger to be allocated. They must be sent to temporary storage—*early baggage storage* (EBS). Figure 3.11 illustrates the system lifecycle of a bag with the mentioned phases.

Given those criteria, the traditional approach for controlling a BHS uses a rather simplified policy of routing totes along static shortest paths. The *static shortest path* is the shortest path of an empty system, but during operation, minor queues are unavoidable, and they lengthen the static shortest routes. In the traditional control, all totes are sent along the static shortest routes, irrespective of the time to their departure, in order to keep the control simple and reliable. A more optimal solution would be to group urgent baggage and clear the route by detouring bags with a distant departure time along less loaded areas.

On top of the basic approach, the control software is fine-tuned against a number of case studies to avoid deadlock situations, but basically it limits the number of active totes in different areas of the system. The fine-tuning process is time consuming and costly for developers; hence, a more general and less system specific solution is one of the ambitions with an agent-based solution.

Naturally, the control of the BHS should try to maximize throughput and capacity of the BHS, which is indirectly linked to the issues of rush bags. Besides that, a number of secondary performance parameters apply as well, such as

Figure 3.11 States of a bag in the BHS

minimizing energy consumption of the motor and lifetime of the equipment—for example, by minimizing the number of start and stops of the elements and avoid quick accelerations.

11.2 Worst-Case Scenario

Apparently from the descriptions given here, there should be opportunities for improvement of the control logic in the BHS, and one might ask why it has not been tried before. It has:

> Still listed as one of the history's top ten worst software scandals are the BHS of Denver airport in Colorado, US. The Denver International Airport was scheduled to open in October 1993, but caused by a non-working BHS the opening of the airport was delayed in 16 months costing $1 million every day. When it finally opened in 1995 it only worked on outbound flights in one of the three terminals, and a backup-system and labour-intensive system was used in the other terminals (Donaldson 2002).

The original plan for the BHS developed and built by BAE was also extremely challenging, even compared to many BHS built today. Instead of moving totes on conveyors, the BHS in Denver is based on more than 4,000 autonomous DCV (destination coded vehicles) running at impressive speeds of up to 32 kph on the 30-km-long rail system. It was a kind of agent-based system with many computers coordinating the tasks, but the first serious trouble was caused by the overloaded 10M-bit ethernet. Also, the optimistic plan of loading and unloading DCVs while running caused DCVs to collide, throwing baggage of the DCVs and sometimes damaging baggage. The original plan even called for transferring baggage from one running DCV into another, whereas many systems today still stop a tote or DCV before unloading, even at stationary discharging points.

11.3 Agent Design

The *elements* are the building blocks of the BHS and from an intuitive point of view are the potential candidates for agents in the system because all actions of the system are performed by the elements. The elements are the module the BHS consists of, mostly conveyor module, which varies 2–9 meters in length (some straight, some curved), but they could also be a module that can tilt or that split. But they are part of the lanes where the baggage moves around in these barrels, called totes.

The applied approach concentrates on the reasoning part of agents and their interaction, from a macro-level perspective. An alternative approach would be to consider the totes as "consumer" agents and the BHS as a collection of "producer" agents, where the BHS can solve the tasks that the totes want to have performed—bringing the tote to the destination. In principle, a tote could then negotiate its way through the system, and if the timing was urgent, the bags would be willing to pay a higher price than nonurgent bags.

Such an approach often leads to other complications, such as communication overhead and complex agent management (Brennan and Norrie 2003). Because the BHS generally consists of pathways of first-in-first-out (FIFO) queuing conveyors with little and often no possibilities of overtaking each other, it is more appropriate to design the agents around the flow of the BHS, which makes the elements the potential candidates for agents. The element agents should then coordinate their activities to optimize system performance and should therefore be considered as collaborative agents, rather than competitive agents.

11.4 Toploader

The input facilities of the BHS are called *toploaders*, as they drop bags into the totes from a conventional conveyor belt (see Figure 3.12). Before the bag is *inducted* into the tote, it passes a scanner that reads the ID tag and destination, so the control system has exact tracking of the bag at all time.

Identity and destination of the bag are unknown until the bag passes the scanner at the toploader. The scanning initializes routing of the tote, but the short time leaves no option for global optimized planning of all current totes with replanning for every new arrival.

Basically, the task of the toploader could be decomposed into several steps. Scanning the bag, which happens automatically, has no direct impact on the

Figure 3.12 A toploader, where bags arrive on a traditional conveyor belt

control. The toploader initiates the journey of the tote on the BHS. A destination (discharger) must be set for the tote. In order to increase capacity, several dischargers are often allocated to the same flight destination.[1] Therefore, the toploader agents initiate a negotiation with the possible dischargers to find the best-suited discharger. The evaluation of the proposals from the dischargers is not trivially chosen as the lowest offer, but weighted with the current route length to the dischargers, which the toploader requests from a route agent—a mediator agent with a global focus on the dynamic route lengths of the BHS.

The toploader can take two different approaches for routing the tote:

1. *Routing by static shortest path*. After the toploader has decided on the discharger, it could instruct all diverting elements along the route to direct that specific tote along the shortest path. Then the agent system would, in principle, work as the traditional control system by sending all totes along predefined static shortest routes.[2]
2. *Routing on the way*. Instead of planning the entire route through the BHS, the toploader could just send the tote to the next decision point along the shortest route. This is a more dynamical and flexible approach, as the tote can be rerouted at a decision point if the route conditions have changed—perhaps another route has become the dynamical shortest one, or the preferred discharging point have changed.

More formally, the principal tasks of the toploader can be illustrated as the diagram in Figure 3.13, but it hides the advanced decision logic between the state changes and message interactions:

- *Straight elements*. Most of the elements of a BHS are naturally straight or curved elements (conveyor lanes) that connect the nodes of the routing graph. Straight or curved elements are not considered as agents in our current design, because mechanically they will always forward a tote to the next element if it free; thus, there are no decisions to be made. In principle, the speed of each element could be adjusted to give a more smooth flow and avoid queuing, so one could argue that these decisions should be taken by the element itself. In the current setup, it would generate an enormous communication overhead, because each element should be notified individually and the agents should be very responsive to change the speed in order to gain anything from speed adjustments.
- *Diverters*. Divert elements (Figure 3.14) become the first natural decision points on a route. A diverter splits a conveyor lane into two, either a left or

[1] Due to the stopping of totes while unloading, the discharger has a lower line capacity than straight elements.

[2] In the researched BHS the decision between the alternative dischargers would also be predefined in the conventional control. The BHS is built in layers to minimize cost and maximize space utilization, and alternative dischargers are always split on different layers. The control system would try to avoid switching layers.

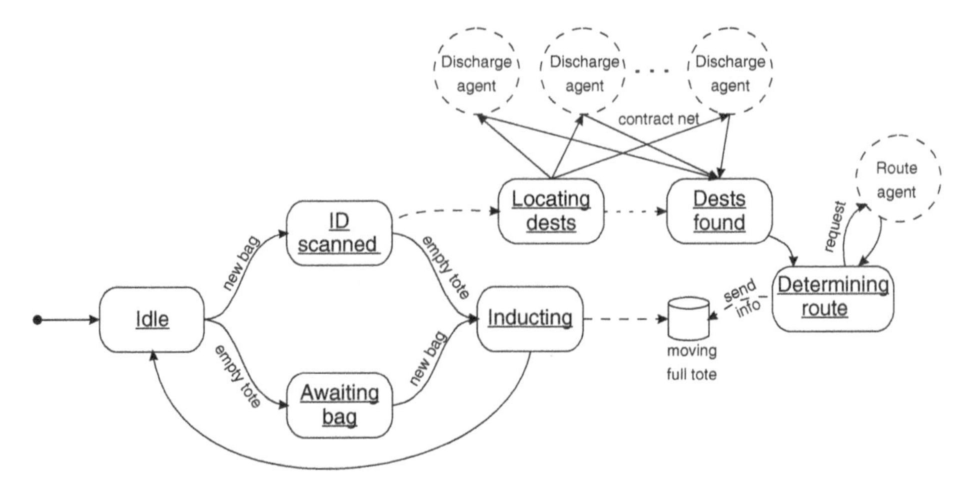

Figure 3.13 The principal tasks of a toploader

right turn and straight ahead. Lifts and so-called cross-transfers are special editions of diverters. The cross-transfer allows the tote to be forwarded in all four directions.

In respect to the strategies described here, the diverter could either just forward the tote in the direction determined by the toploader, or it should reconsider alternative routes by restarting the negotiation process with dischargers and requesting updated information on dynamic route lengths. A

Figure 3.14 A diverter element with an empty oversize tote

diverter should be concerned about the relevancy of reconsidering the route for a tote, because in many cases there is only one possible direction at a given diverter for a given tote.

So the decision logic, rather, is identical to the dynamic routing principle of toploaders, but diverters should fine-tune their decisions according to the local environment in which they are situated. In other words, a strong influence on the decision logic of the diverter is based on its position in the routing graph.

Similar to the toploader, the principal tasks of the diverter can be illustrated by a state diagram, shown in Figure 3.15.

- *Mergers*. Mergers are the opposite of diverters, as they merge two lanes. Traditionally, mergers are not controlled, as there are no alternatives to continuing on the single lane ahead, and the merger simply alters between taking one tote from either input lane, if both are occupied.

  Obviously, more intelligent decisions could be considered than just switching between the input lanes, which is the argument for applying agents to the merger elements. The ratio between merging totes from the input lanes should be determined by the aggregated data of the totes in either of the two lanes (e.g., if the number of urgent totes waiting to be merged is higher in one lane, then that lane should be given higher priority). Also waiting totes in one lane could have greater impact on the overall system performance, if a queue of totes in one lane is more likely to block other routes behind that point.

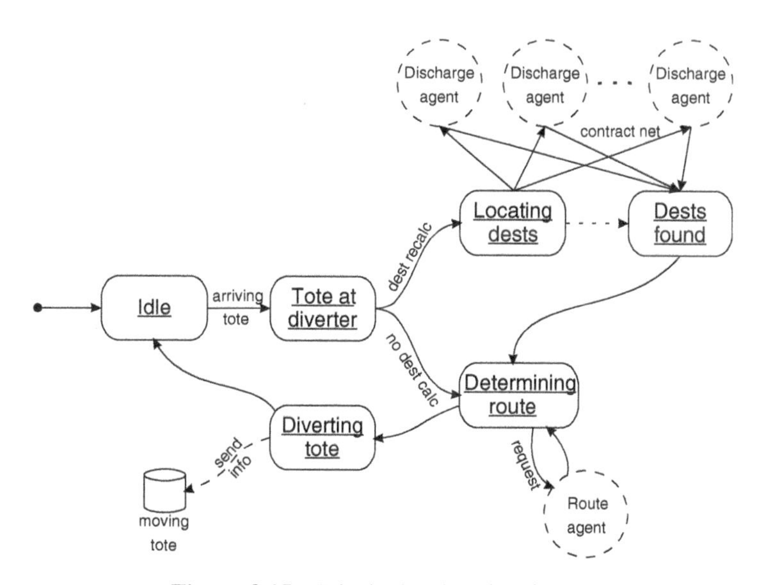

Figure 3.15 Principal tasks of a diverter

Figure 3.16 A discharger element that can tilt the tote, so bags slide onto the conveyor belt

- *Dischargers*. Dischargers (Figure 3.16) unload bags from the totes. When bags are discharged they fall onto carrousels similar to those at baggage claim and are drown to the plane in small wagon trains.

  Besides being involved in the negotiation process described for the toploaders, the task of the discharger could seem rather simple—just tilt the tote—but a discharger also has to take care of the empty totes. Some BHSs have a separate conveyor system for the empty totes, but many systems, including the researched BHS, use the same lanes for routing the empty totes back to the tote stackers at the toploaders.

  The task of routing empty totes is similar to routing full totes at the toploaders, but is actually much more complex, due to a number of considerations that must be taken into account:

  - The number of destinations (tote stackers) is larger than alternative dischargers for full totes. The number of tote stackers often matches the number of toploaders, which is 12, in our case.
  - Especially in the input area, empty totes are mixed with full totes, and the area could easily get overloaded and blocked.
  - During peak times, some empty totes should be sent to temporary storage in the EBS, which is far from the toploaders, and then released when the load on the system is lower.
  - If a stacker runs empty, no totes will be available at the toploaders for new bags.

- The distance to the stackers is more appropriate to return the empty tote to a stacker nearby than sending it half way through the system.

  All these factors could be considered in the decision logic of the agent (e.g., by using some fuzzy set logic). The principal tasks of the discharger are illustrated by the state diagram in Figure 3.17.

- *EBS elements*. Early baggage storage elements (Figure 3.18), or EBS for short, are temporary storage elements for totes with bags for which a discharger has not been allocated yet, as already described when defining the concept of rush bags.

  EBS is a complete research area in itself regarding optimization of EBS elements, as totes are stored in lanes, which are released into the system again. Planning and coordinating the totes in different lanes is not a simple task, but was not given further attention in the project.

11.5 Agent Interactions and Ontology

The agent interactions are based on the elements responsibility and participation in the function of the BHS, as described in the previous section. To give an example of a delegate or mediator used in the system for a federation among the agents, the RouteAgent is described.

There is a balance between giving agents detailed information about the environment and maintaining an internal world model, or let agents query the environment about information when required.

In theory, the interagent negotiations could be used to generate all information to route totes around in the system, but that would generate too much overhead

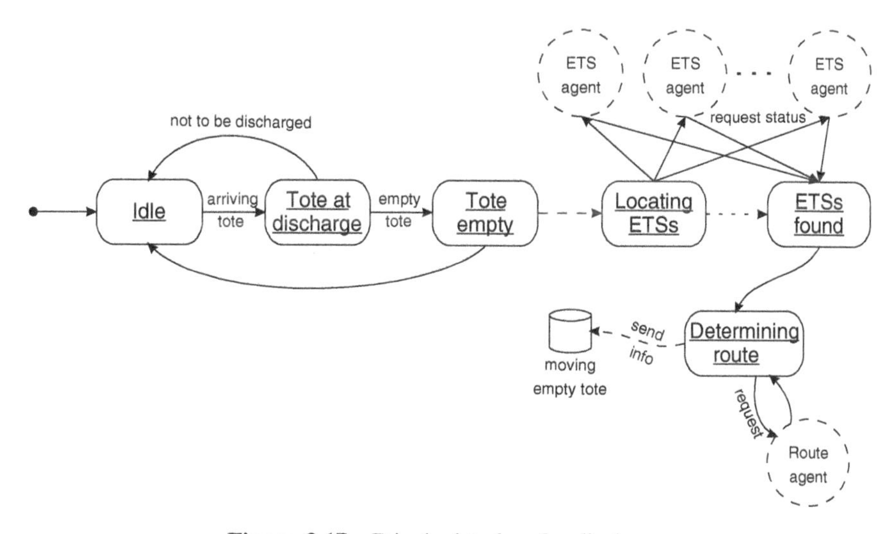

Figure 3.17 Principal tasks of a discharger

Figure 3.18 EBS elements, here storing a line of empty totes

and complicate the simple routing principles. Instead, agents can be assisted by a mediator agent that collects aggregated information for the entire system. In the initialization process, the RouteAgent generates all possible routes in the system by building up a graph for the BHS with nodes corresponding to the element agents. During the operation, it constantly monitors traffic on all edges of the graph and update the weights in the graph, so dynamic shortest paths can be calculated using classic Dijkstra for dynamic shortest path calculations (Dijkstra 1959).

Following the FIPA *query-ref* communication act, element agents can request routes to a given destination packed in a referential expression of the query message. The referential expression is composed using an ontology that has been defined for the BHS domain, which extends and follows the structure of the FIPA-SL. The RouteAgent understand two concepts of the ontology, *RouteBetween* and *LineBetween*:

1. *RouteBetween* is the concept used, when agents are interested in full or parts of a route, but only with a granularity of finding other element agents along the path—only information on nodes of the graph are returned.
2. *LineBetween* is the fine-grained concept providing all details about a conveyor line of connected elements in the BHS—information about edges between two given connected nodes.

To give an example of the generality embedded in ontology-based messages, a query to the RouteAgent could contain the following abstract referential expressions:

```
(iota
 :Variable (Variable :Name x :ValueType set)
 :Proposition (routeBetween
  :origin (element :elementID DFB01.TLA001)
  :destination (element :elementID DLA02.DIA023)
  :viaPoints (Variable :Name x :ValueType set)
  :numNodes 0
 )
)
```

It is abstract because it contains the variable **x** that must be replaced by the responder in a response to the query. In this case, the responder is a set of points (identities of element agents between the given origin and destination). The predicate iota is just one of three from the FIPA-SL specification, which means exactly one object fulfills the expression, whereas the other predicates, *any* and *all*, would return any or all routes between the origin and destination, respectively.

11.6 Internal Agent Reasoning

This section will present internal agent reasoning principles to optimize the flow in the BHS in different ways to meet some of the performance parameters. Deep reasoning and long-term goals are not currently pursued in the strategies, due to the flow speed and high number of totes in the system. Instead, the intensions behind the strategies are to optimize the situation for more than a single tote or forthcoming actions.

Even though the agent design does not strictly follow the BDI architecture, the behavior of the agents follows the same principle, with agents constantly monitoring the environment, and it will change its actions according to the goal it is design to achieve based on the current state of the environment.

Three different deliberate behaviors of agent will be described, which are part of both necessary routing and optimizing strategies for the BHS. The deliberate behaviors are included to exemplify how agents can have very diverse internal reasoning, which would be very hard to combine in a central solution. Intuitively, they are much easier to understand and implement when taking the perspective a single agent, its environment, and the agents it has to collaborate with.

11.6.1 Returning Empty Totes

As already explained, the task of dischargers is more complicated than just emptying the tote. The tote continues on the conveyors and should be routed back to tote stackers located at the input facilities. The most important factor that influences the decision of the destination for the empty tote is the full status of the tote stackers. However, the distance to the tote stackers should also be considered. There is no reason to send it to the other end of the system, if a stacker is nearby, unless the other is empty.

Each stacker monitors its full status as a simple ratio between the current and maximum number of totes in the stacker. By a standard hedge (Negnevitsky 2005) the ratio is converted into a priority s_i for requesting extra totes:

$$s_i = \begin{cases} 2r_i^2 & 0 \leq r_i < 1/2 \\ 1 - 2(1 - r_i)^2 & 1/2 \leq r_i \leq 1 \end{cases}$$

where r_i is the full-ratio for the ith stacker. A plot of the function is shown in Figure 3.19.

The priority is used to scale the dynamic route length to each tote stacker, so a nearly empty stacker will have a very short route length or value in the decision, whereas a full stacker will have its full route length:

$$v_i = d_i \times s_i$$

where d_i is the dynamic distance (requested from the RouteAgent) to the stacker from the decision point.

11.6.2 Overtaking Urgent Bags

Consider a typical layout of a discharging area in Figure 3.20. The bottom lane is a fast-forward transport line, the middle a slower lane with the dischargers and the upper lane is the return path. A diverter (in the bottom lane) has the option to detour nonurgent to the middle lane to give way for urgent baggage in the transport line, but with no queues in the system all totes should follow the shortest path. When the routes merge again at the mergers in the middle lane,

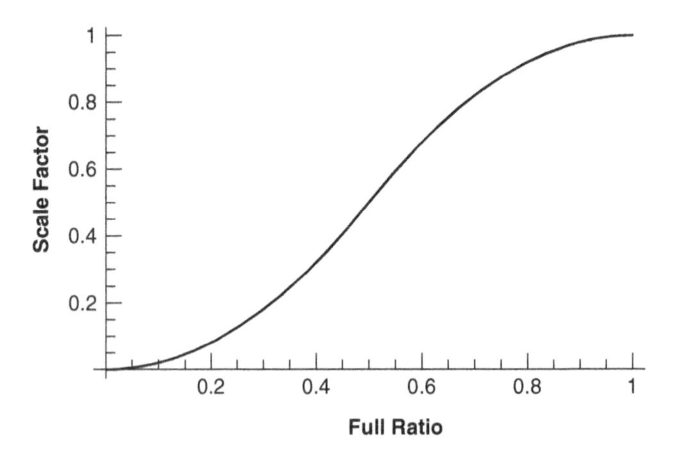

Figure 3.19 Plot of ETS priority function

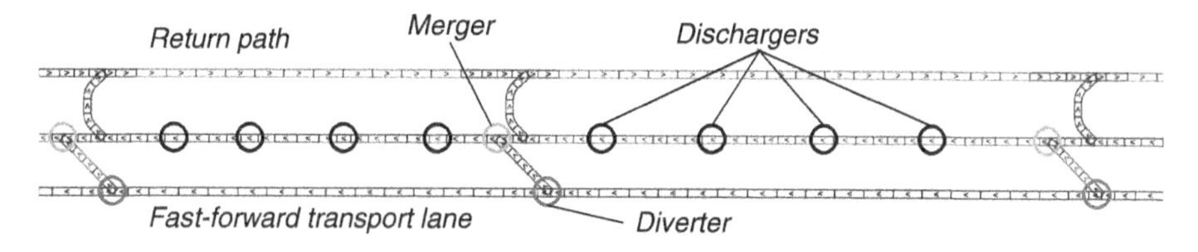

Figure 3.20 Area of the BHS layout with indication of diverters, mergers, and dischargers

that lane will give higher priority to totes from the merging leg with the most urgent baggage.

Urgency is a constructed function, which gives high priority to urgent totes and negative priority to totes where remaining time to departure exceeds a threshold.

$$u_j = \begin{cases} \dfrac{1}{{t_j}^2} & t_j < U_T \\ \dfrac{1}{(U_{\max} - U_T)^2}(-{t_j}^2 + 2U_T t_j - {U_T}^2) & t_j \geq U_T \end{cases}$$

where $U_{\max}$ is the full window size of the allocated discharger. If the tote's remaining time exceeds this value, it should go to EBS. U_T is the threshold value, which is set to 20 minutes, as no tote should be considered urgent if it has more than 20 minutes left before the discharger closes.[3] t_j is the remaining time for the jth tote. The graph is plotted in Figure 3.21.

The urgency factor is converted to a scale factor for the dynamic route lengths of alternatives routes. Then the principle of simple modification of the route

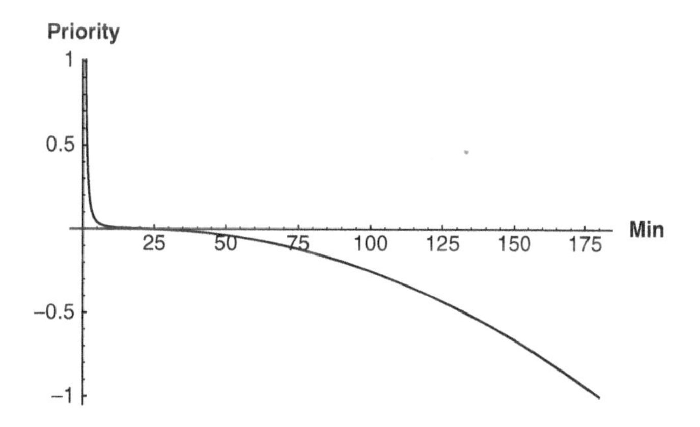

Figure 3.21 Urgency function for totes

[3] When the discharger closes the tote becomes a rush bag, but the threshold of 20 minutes is independent of the 20 minutes time limit for rush bags, so in total a tote is considered nonurgent if it has more than 40 minutes left to departure.

lengths can be used here as well:

$$s_j = \begin{cases} (1-u_j)(1+v_{k+1}) & u_j < 0 \quad \text{(nonurgent tote)} \\ (1-u_j)(1-v_{k+1}) & u_j \geq 0 \quad \text{(urgent tote)} \end{cases}$$

where v_{k+1} is the aggregated urgency value for the next decision point along the route, which is requested in a communication act (FIPA *request-ref*) to the divert agent. The formula secures that urgent totes will group along the shortest route (as v_{k+1} is close to 1), whereas nonurgent totes are punished along the detour. If there are no queues on the routes the v_{k+1} is 0, and the scale factor has no effect.

The mergers in the middle lane simply give higher priority to input lanes with more urgent totes. The ratio between the aggregated urgency factors of the input lanes becomes the ratio for merging totes from the input lanes.

11.6.3 Saturation Management

Another important strategy is trying to avoid queues by minimizing the load on the system in critical areas. Consider slow-starting queues of cars at an intersection when the light turns green. Acceleration ramps and reaction times relative to drivers ahead accumulate to long delays in traffic queues, even though, in theory, all drivers should be able to accelerate synchronously (no reaction time).

The same problem arises in the BHS, where reaction times correspond to the delay of the element head reporting clear.[4] These matters result in the characteristics of the work in progress against capacity curve (WIPAC), which is further described in Kragh (1990), who states that the capacity of a system goes down dramatically if the load on the system exceeds a certain threshold value, as indicated in the Figure 3.22.

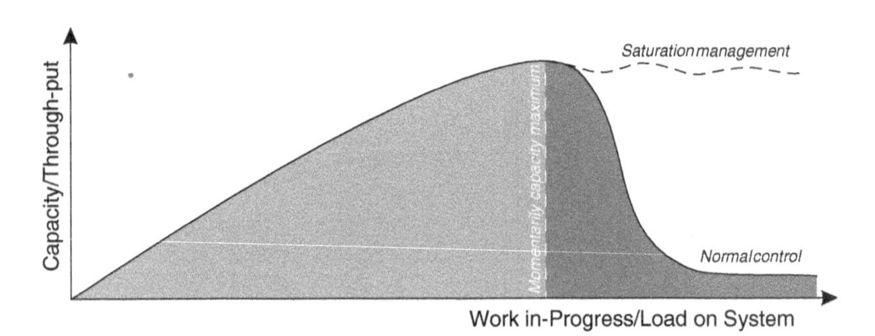

Figure 3.22 Theoretical WIPAC

[4] In the mechanical setup of the BHS, a tote can only be forwarded from one conveyor element to the next element if that element is clear. A synchronized row of totes can then pass at full speed, from one element to another. In queue situations, acceleration ramps delays each element.

The curve is dynamical, due to the various and changing load on the system, and the maximum cannot be calculated exactly. Thus, the strategy is to quickly respond to minor observations, which indicate that the maximum has been reached, and then block new inputs to the area. This approach is called *saturation management*, where the toploaders will be blocked if the routes from the toploader are overloaded.

Queues close to the toploader are most critical, as the toploaders have great impact on filling up those queues, whereas the parts of the route far from the toploader could easily have been resolved before the new totes arrive. Instead of blocking the toploader completely, it can just slow down the release of new totes using the following fraction of full speed:

$$v_t = \frac{\sum_i w_i q_i}{\sum_i w_i} = \frac{\sum_i \frac{\alpha}{d_i} q_i}{\sum_i \frac{\alpha}{d_i}}$$

where v_t is the full speed of the toploader, $w_i$is weight of the queue statuses, q_i, along the routes. The weight is given by a fitted coefficient, α, and the distance from the toploader d_i. Queue statuses, q_i, are always a number between 0 and 1, where 1 indicates no queue.

The effect of the saturation management strategy is clearly documented by the graph in Figure 3.23. Thus, the decision taken by the toploader agent is highly dependent on the current configuration of the environment around the toploader.

12 CASE STUDY 2: MATERIAL HANDLING IN AN ANODIZATION SYSTEM

The second case is a material handling system that moves bars of items between different chemical baths. Each bar has its own recipe to process the system, and system scheduling is modeled by simple reactive agents that influence each other through their actions on the environment. Therefore, there is no direct negotiation between the agents; it is more a matter of coordinating their activities.

Timetabling of classes is a classical constraint satisfaction problem, which is known to be hard or even NP-complete for large schools or universities. But

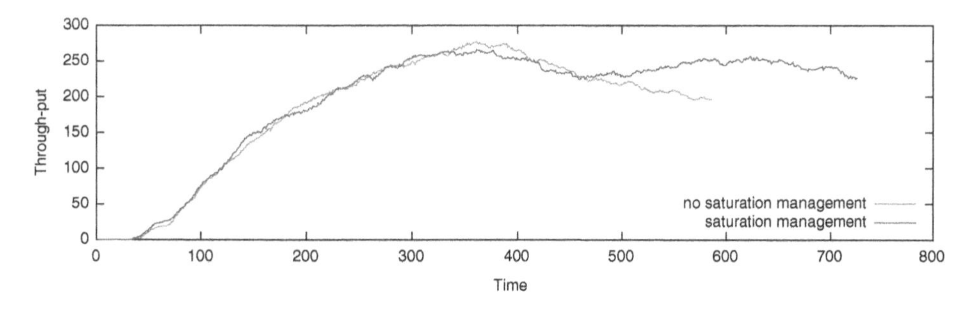

Figure 3.23 Result of a test scenario with and without the saturation management strategy

imagine the increase in complexity if duration of class sessions were allowed to vary dynamically in length. The argument could be that to take full advantage of the resources (teachers and classrooms), teachers should only stay as long as required for the students to understand the topic, but bounded by a minimum and maximum timeframe.

This case study is based on a project conducted in collaboration with Denmark's most well-known manufacturer of high-end audio and video products. The products are respected worldwide for their extremely high-quality finish and design, and the investigated production facility is the process that gives the surface of the product the high-quality finish. The process is known as an *anodization process* that increases the corrosion resistance of aluminum, but coloring of the surface is also part of the process.

In a generalized and simplified form, the problem could be described as a number of chemical baths, which the items have to visit according to a prescribed recipe. Besides containing information about which baths to visit and in what order, the recipe also gives an allowed time frame for the item to stay in each bath. Items are grouped on bars with the same recipe, but a mix of different bars (that is, different recipes) could be processed at the same time in the production system.

The system consists of about 50 baths, and a typical recipe would have roughly 15 to 25 baths to visit. Even though all recipes do not have to visit all kind of baths, there is still room for additional baths of the same type to overcome bottlenecks, as the processing times in the baths types vary a lot. Thus, the recipe contains only bath types, not bath number, and it is the task of the control software to allocate a specific bath among duplets for every bar.

Three slightly overlapping cranes move the bars from one location to another in the array of baths. Here, a simplified notion for the movements will be used, but in practice, they are more complex than that, because moving between some specific baths includes subprocesses such as rinsing the bar of items, opening and/or closing the lid of a bath, and so on, but it comes down to an estimated travel time of moving a bar from bath u_j to bath u_{j+k}. In general, the cranes are not considered to be a bottleneck in the production system, as they handle the tasks quite sufficiently.

Apart from the baths and cranes, an important part of the system is the input buffer, where typically around 30 bars are waiting to be processed. This also is an important focus point for the control software, because choosing the best bar to fit the current configuration is the key to optimizing throughput. A general overview of the system is presented in Figure 3.24, where the C are the cranes with a domain (how far they can move) and the U are all the chemical baths into which the bars can be placed.

At first sight, the problem described seems to be a candidate for classic optimization and scheduling principles, but as already mentioned, the allowed time

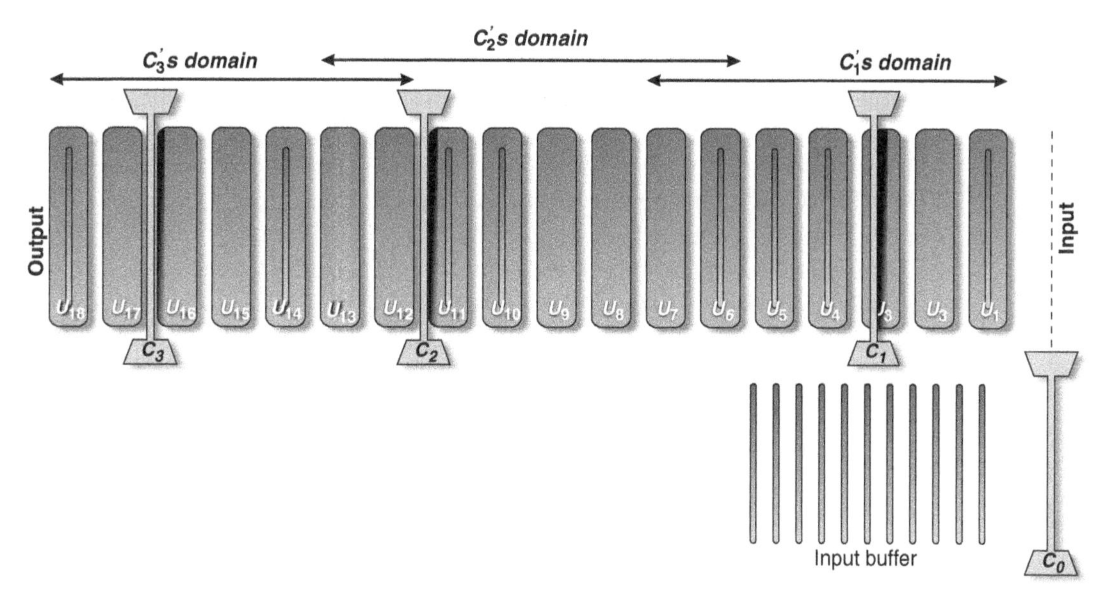

Figure 3.24 Generalized overview of the system for the anodization process, with bars moving from right to left

frame for processing each step of the recipe complicates the task.[5] Another issue is the dynamical production environment, which has great impact on the system's ability to recover and finish the current bars while running as best as possible under partial breakdowns. Examples of unpredictable error conditions could be that the temperature of a bath is too low and must be heated before being available again, cranes break down, the liquid level of a bath is too low, or orders are too rapid. An agent-based approach must focus on the dynamics and be able to recover or continue as best as possible under such conditions.

The problem is classic—the throughput of a system should optimize the flow between subprocesses and handle the inflow process correctly to best utilize the system.

In abstract terms, there exists a number of tasks, q_i for $i = 1, 2, \ldots, n$, with k_i subtasks[6] $q_{i,1}, q_{i,2}, \ldots, q_{i,ki}$. The subtasks are interconnected, and the order cannot be changed. Tasks should be handled as visits to processing stations—determined by the recipe.

12.1 PACO Approach

PACO is a contraction of *coordinated patterns* (Demazeau 1991) and takes a simple approach of designing the agents. PACO focuses on reactive agents situated

[5] Only minimum and maximum times are given for each step of the recipe.

[6] Note that the number of subtasks might be different for each taskgroup.

in an environment, where all agents are considered as partial solutions of a global problem (Gufflet and Demazeau 2004).

Interactions between the agents and the environment are generally applied and modeled as forces, and by giving the agents a mass, they will—at least from a conceptual point of view—have both velocity and acceleration, which is valuable when adjusting the priorities between interactions. The applied forces are springlike forces, which reduce the risk of oscillating interactions but also secure that the system will converge to an equilibrium state at some time in the coordination process between all agents.

The PACO paradigm states that agents are purely reactive; thus, they do not hold an updated internal representation of themselves, other agents, or the environment, so they have to respond to all change of the environment in which they are situated. This general idea suits the researched case very well, as agents after an initialization process will hold some kind of plan for handling the current set of bars in the system. Whenever a new bar is introduced, or some kind of unforeseen or expected events happen within the system, such as when a crane breaks down or a bath needs cleaning, it is just a new stimulus to the agents of the control system, and they will start searching for a new equilibrium state through their interactions.

Each agent under the PACO paradigm is defined by three fields, which divides the agent model in three coherent components:

1. *Perception field* determine what the agent can perceive about its environment.
2. *Communication field* determine which agents an agent can interact with.
3. *Action field* determine the space in which an agent can perform its actions.

From a system point of view, the PACO paradigm also splits the system into conceptual parts, which follows the Vowels formalism (Silva and Demazeau 2002) explained earlier.

12.2 Agent Design

This section describes and discusses how the PACO approach under the Vowels formalism has been applied to the researched anodization system. The following subsections cover each part of the method:

- *Environment*. Before agents can be created and assigned with goals and behaviors, there must be an environment for them to exist in. In this case, the environment is the baths and cranes. The environment is modeled as passive resources, which the agents can ask about their status and book for a given time. Baths are accessed through a bath controller, which makes baths of the same type look like only one bath in the software that is capable of containing more than one bar at a time. These baths can be asked about free space in a given direction by an agent or about whether a free time slot of a required time frame exists in a specific period. If the space is occupied, then the bath can tell which agent is blocking. A bath has no possibility to prioritize or assign time to individual agents, or to push or cancel time already assigned. It is the responsibility of the agents to fight for time slots themselves.

- *Agents*. The recipe for each bar of items is split into a number of agents. One agent is created for each step of the recipe, and all agents made from the recipe form a group. An agent is born with some knowledge, as it knows which kind of bath it must visit, it holds the allowed minimum and maximum time to stay in the bath, and it knows its predecessor and successor agent of the group. It does not know the rest of the agents in the group and it has no way of communicating with them. Thus, the scope of the agent within its group is rather limited, which simplifies the interaction model.

To succeed, an agent must visit a bath of the right type, but not necessarily at the right time. The agent has a size equal to the time slot it occupies in the bath. Therefore by its representation, agents can be seen as physical manifestations of the problem in focus. Two bars q_i and q_j, split up into two groups of agents, $q_{i,1}, q_{i,2}, \ldots, q_{i,1}, q_{i,2}, \ldots, q_{i,n}$, and $q_{j,1}, q_{j,2}, \ldots, q_{j,m}$ added to the model in random places it could look like Figure 3.25.

12.2.1 Perception, Communication, and Action Fields

As stated earlier, the PACO paradigm defines three delimited fields of how PACO agents experience the world: the perception, communication, and action fields.

- *The perception field* consists of the predecessor of the bar, as the movement of that agent affects the forces (described later in this section) applied to an agent. Furthermore, the agents above and below (in the time domain) that want to visit the same bath are also observed, to avoid overlap of agents in the same bath.

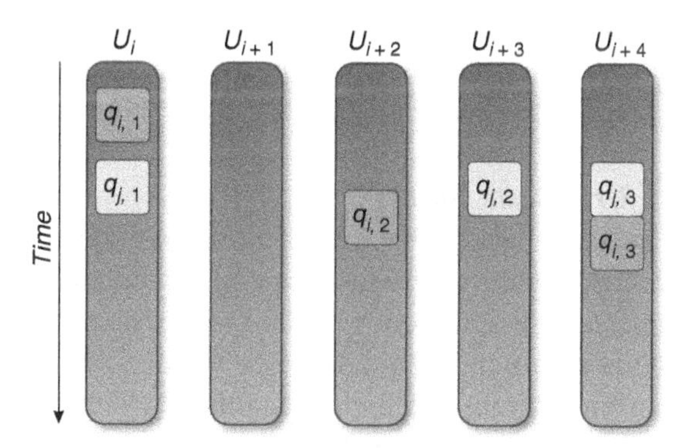

Figure 3.25 Three agents from each agent group occupying timeslots in the baths

- *The communication field* solely consists of the predecessor, as it should be notified if the agent could meet its goals.
- *The action field* consists of the baths of the requested type, and organization ensures that an agent only sees one particular bath, even if the bath type is duplicated in the system.

12.2.2 Agent Goals

An agent has two main goals:

1. Go in the right bath.
2. Stay close to the predecessor agent of its group.

When both goals are satisfied, for all agents of a group, the bar represented by the group has a valid way to be processed by the system. Furthermore, an agent has three constraints:

1. Keep distance to both the min and max time.
2. Help the successor to stay close.
3. Help other agents in same bath type to fulfill goals.

Constraints are added to make agents cooperate with others in fulfilling their goals, too. When agents from two groups share interests to the same time slot for a given bath, they have to be able to negotiate which will win the time slot.

Therefore, an extra type of agent is introduced—an observer. For each group of agents, one observer agent monitors the movements and how satisfied agents are in general. Information withdrawn from this observer is used when solving conflicts.

12.3 Interactions

Agents move around in the virtual world in discrete steps. They calculate a force vector v as responses to input/output from the three fields. Each discrete time step has two parts. First, all agents gets a parallel chance to decide which way to go and at what speed. Hereafter, they get the chance to move themselves. In this moving step, they will try to move in the direction and distance specified by v, within the space allowed by their action field.

12.3.1 Basic Forces

The most basic behaviors of the agents come from their primary goals and are modeled with two forces: a spring force and a gravity force. The spring force, F_s, represents the attraction to the predecessor, if any, and attracts the agent towards the point where the predecessor's time slot of the previous bath ends, so the bar can move from one bath to another, which is a criteria for a plan to be valid.

In general, a spring force is denoted: $F_s = -kx$, where k is the spring constant and x the distance, so in this case: $F_s = -k_{\text{parent}}(x - x_p)$, where k_{parent} is a

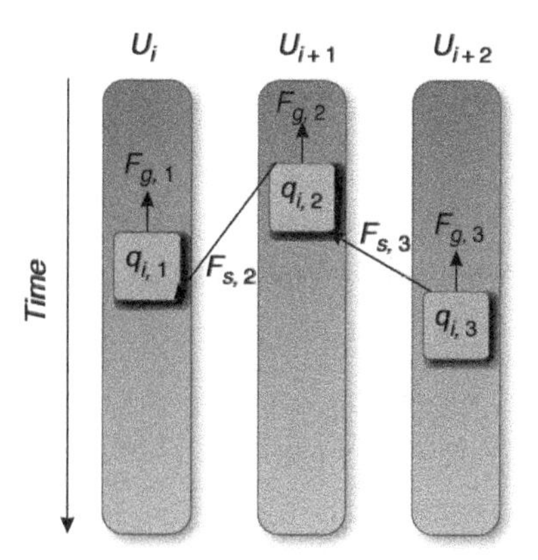

Figure 3.26 Basic forces for the agents

static constant, x is the position of the agent and x_p is the ending point of the predecessor.

The second force, the gravitational force, tries to pull the agent up. *Up* in the virtual model represents beginning of time in the real world, as shown in Figure 3.26. The gravitational force is given by $F_g = mg$, where m is the mass of the agent and g the gravitational acceleration. With the mass of all agents being the same, $F_g = k_g$, where k_g is a static constant force vector. This gravitational force is only applied to an agent when it is floating freely. If the agent is in contact with another, in the direction pulled by the force, the counterforce from the contact will cancel out the gravity. The total force is denoted F_t:

$$F_t = F_s + F_g$$

With only these two simple forces, a set of agents can be added that can align themselves and thereby make a valid schedule of how to be processed by the system. See Figure 3.26, where U_{i+1} is the next bath the q_i bar has to visit, the F_s are the spring forces between these agents, and the F_g are the gravitational forces between these agents.

The spring force serves to compact the plan of an agent group in order to minimize the total processing time of a bar, whereas the gravity force works to compact the entire plan for all bars in order to maximize utilization of all baths.

12.3.2 Organizations

To make the interaction between the agents more flexible, six social laws are introduced:

Law 1

If there is a certain amount of free space around the agent, increase size to

$$T_{\text{current}} = T_{\min} + X(T_{\max} - T_{\min})$$

where X is a static constant and $T_{current}$, T_{min}, and T_{max} are the current, minimum, and maximum time slots of the agent.

Law 2

If another agent, using the same bath, approaches within a given distance, then shrink the current size until T_{min} plus a given margin is reached.

Law 3

If the successor is unable to reach its second goal, then stepwise increase $T_{current}$ until T_{max} is reached.

If an agent needs to go in a direction blocked by other agents, it should be able jump over, push, or switch places with one of the blocking agents, as illustrated in Figure 3.27. For this purpose, the remaining social laws apply. They are respected when agent A wants to go in a direction blocked by agent B.

Law 4

If F_t for A is greater than the current size of B, and a time slot of at least A_{min}, is available between the end of B and the length of F_t, A jumps to the other side of B, without notice.

Law 5

If there is no room for A on the other side of B, but B is trying to move in the opposite direction, and if the size of F_t for A is greater than half of B_{min}, then they switch places.

Law 6

If neither of two previous laws applies, but A still wants some or all of the time slot assigned to B, then A starts a negotiation based on the general satisfaction of groups A and B. If A wins this negotiation, B is pushed away; otherwise they both will have to stay.

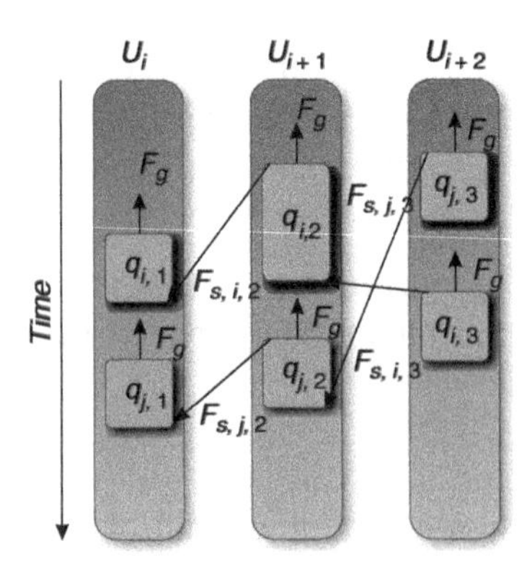

Figure 3.27 A conflict between two agent groups

With this set of forces and laws for solving problems, a decent number of bars should be able to be split up into PACO agents and added to the system. Hereafter, they will align themselves in a roughly optimal way or stay in motion, trying to seek their goals.

Apparently, the real challenges of applying the PACO paradigm to an agent-based control system like the one in this case study is designing and fitting the forces used for interactions.

The agent group, which spawns from the creation of agents for a single bar of items, is not modeled or implemented as a sole entity in the system. Thus, no overall goal or intensions of the group can be directly implemented, but must be realized through the aggregation of subgoals met by the agents within the group. The tension appearing inside a group due to the spring forces of the agents can, to some degree, lead to competitions among agents within a group, but the social laws make it easier for the system to reach an equilibrium and dampen the interagent tensions. Particularly, laws 1 and 3 are added to cope with these side effects of the basic forces. Law 1 simplifies the process of attraction and stabilizes the movements of a successor agent to its predecessor, due to the expansion of the current time slot for an agent in a bath, if it is too hard to pack the schedule for a bar tighter. Note that the plan for each bar at the end must form a consecutive sequence of visits to baths as the cranes move bars from bath to bath, because the system has no spare slots that temporarily can hold a bar. Law 3 more directly compacts the plan of a group and increases robustness in the coordination process.

Law 2 is important as well, even though it is orthogonal to the agent groups. It adds flexibility by minimizing the slot time requested by an agent. It is not a direct coordination mechanism between agents from different groups, but allows some mutual impact on their actions.

The most challenging part of optimizing the overall plan for the system is to decide when and how conflicts between agents should be solved. No method or measurements exists to validate if a current configuration is optimal or jumps between the agents should be handled. Laws 4 and 5 direct the trivial conflicts to be handled without contracting classic local optimization principles. Law 6 serves to dampen intergroup tensions, especially to avoid oscillating shifts between agents from different groups with interest in the same bath.

13 RESULTS

The agent-based approach for the control system can be tested to see if it can create valid plans for the system, which can be done by measuring a satisfactory rate for an agent group. A fully valid plan would have a 100 percent satisfactory rate, which means that for a given bar, all visits to baths in the recipes comply with the minimum and maximum time frames and that moves between two

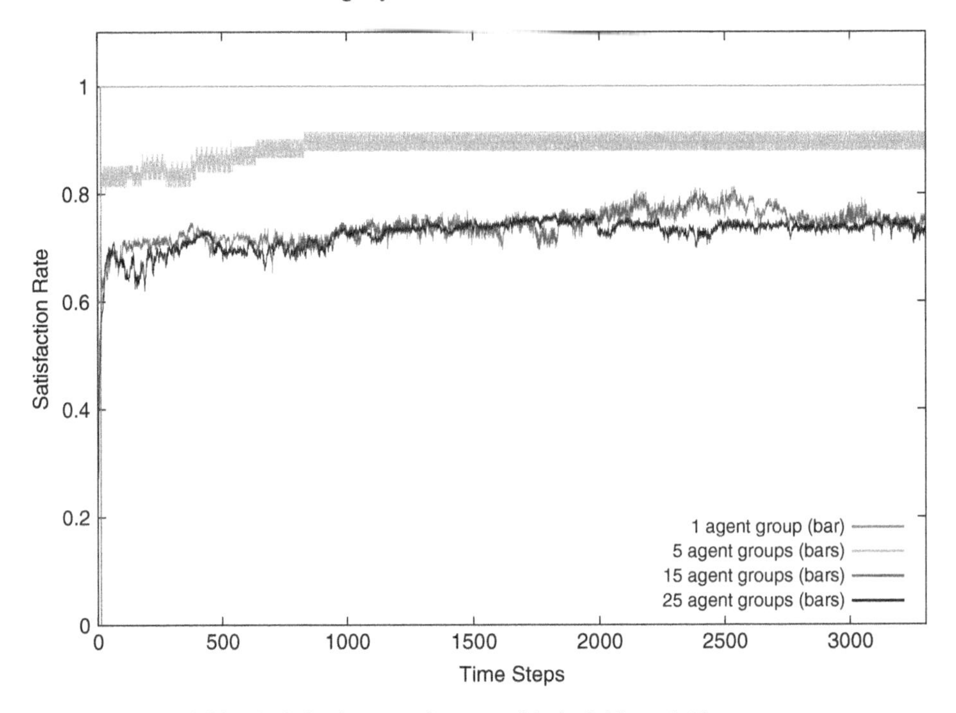

Figure 3.28 Satisfaction rate for test with 1, 5 15, and 25 agent groups

consecutive visits are connected with no glue time (the extra time added to a bath visit for the plan to consist of consecutive visits between the different baths).

Naturally, the computation time is also interesting as a measure for the dynamic reactiveness of the approach, but as the results show in Figure 3.28, the agents relative quickly move to a rather stable level of their satisfaction rate.

Valid plans are not met in all scenarios, so in order to improve the results, a number of experiments have been conducted, which clearly improve the number of valid plans being generated. Some of those strategies will be explained in this section. On a meta-level, they show one of the real strength for developers to work with agent systems: It is very easy to add or change a behavior that is local in one agent and test whether the performance has improved, without restructuring a central control. These strategies also give the agent a more deliberative behavior, so by introducing such changes, the system would be more a hybrid system than a system of purely reactive agents.

13.1 Active, Sleeping, and Locked Agents

During tests, it has generally been observed that the system could end in a nonconverging situation, where one or more agents oscillate. Thus, a promising approach is to give the agents a state that determines their ability to perceive the environment and how they should react to new stimuli. Three states are obvious for the agents:

1. *Active agents* are agents that observe everything of their fields, and fully accept the input and influence of other agents according to the six laws. In other words, active agents behave as expected from the design section.
2. *Sleeping agents* are agents that no longer possess intentions to move, and that other agents no longer can expect to influence. However, stimuli from the environment and input from other agents can grow so strong that the agent is awakened again.
3. *Locked agents* are agents that no longer are under influence of the environment or other agents, but internally can still decide to unlock and become active again.

The interesting issues are the transitions from state to state. For a sleeping agent, there are two options: The agent itself or its group expects that the agent can improve its position, or the agent is being pushed by the environment. An agent could expect to improve its position if free space above has become available, and not necessarily directly above the agent. This could also happen if a larger chunk of free space has become available earlier in the time domain.

The pressure from other agents can be controlled by a threshold value, so the agent is awakened if the forces applied from other agents are too high, given by the summed force of impact from others:

$$F_{I_O} = \sum_i F_i S_i d_i$$

where F_i is the force from the ith agent that want the position, S_i is the satisfaction rate of the ith agent, and d_i its distance in time to the sleeping agent.

Also, the pressure on an agent from its group can be expressed as a summing force that can break a threshold value and bring the agent awake again:

$$F_{I_G} = \sum_j F_j \frac{1}{d_j}$$

where F_j is the applied force from the jth agent in the group and d_j its distance in steps to the sleeping agent. Given those transitions, an agent could fall asleep if it is has found a steady state and is not in conflict with other agents. It could also fall asleep if its group has found a stable level, but some members are oscillating.

13.2 Predecessor Validation

An agent adjusts its position according to the free space around it, but also under influence of its predecessor's position. Experiments have shown that especially during the initial settling time of an agent group, some agents were strongly influenced by their predecessor agents, which were not very reliable with respect to their final position.

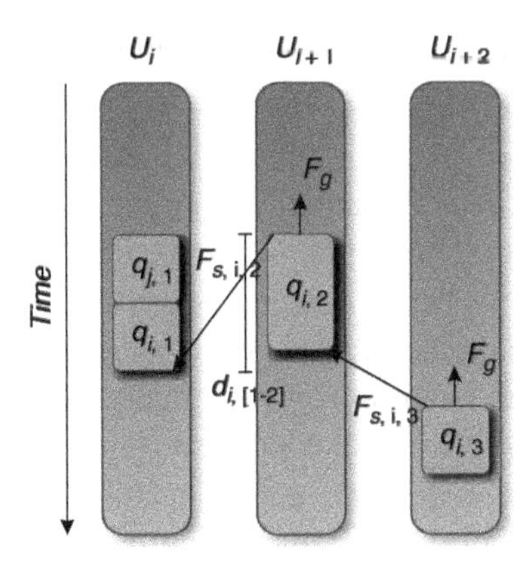

Figure 3.29 Parent validation problem

In the example, agent $q_{i,3}$ would move upward in time due to gravitation and the position of $q_{i,2}$, but it is rather obvious to see that $q_{i,2}$ is not very reliable due to position of $q_{i,1}$ that seems to have settled next to $q_{i,1}$. It is not certain that the parent validation will improve the plan of the system, but it is introduced to dampen oscillations of agents. One way to validate the parent is to look at its position according to its parent, as illustrated by the dimension in Figure 3.29.

Figure 3.30 shows the result of a scenario with 25 bars both with and without the parent validation. As expected, the plan becomes more stable with the parent validation, but it also has a longer settling time as a natural consequence.

13.3 Floating

The last improvement is called *floating agents*, and is best described from Figure 3.31.

According to Figure 3.31, an agent q_i would behave as in the left case (a). The action field of an agent only allows an agent to move in the direction of the resulting force until it is blocked by other agents, in this case q_j, even though the force is larger. By extending the action field to the size of the resulting force and allow the agent to float over another agent q_j many conflicts might be avoided. It is similar to allow the agent to search for a valid position from the bottom of its action field, as illustrated in the right case (b), whereas the agent in (a) searches from the top until it meets a block or the end of the force vector.

Figure 3.32 presents the results of the scenario with and without the floating improvement enabled. There might be more fluctuations with floating enabled, but as expected, it dramatically improves not only the settling time but also the overall satisfaction level, as the agents avoid many conflict caused by tensions between agents.

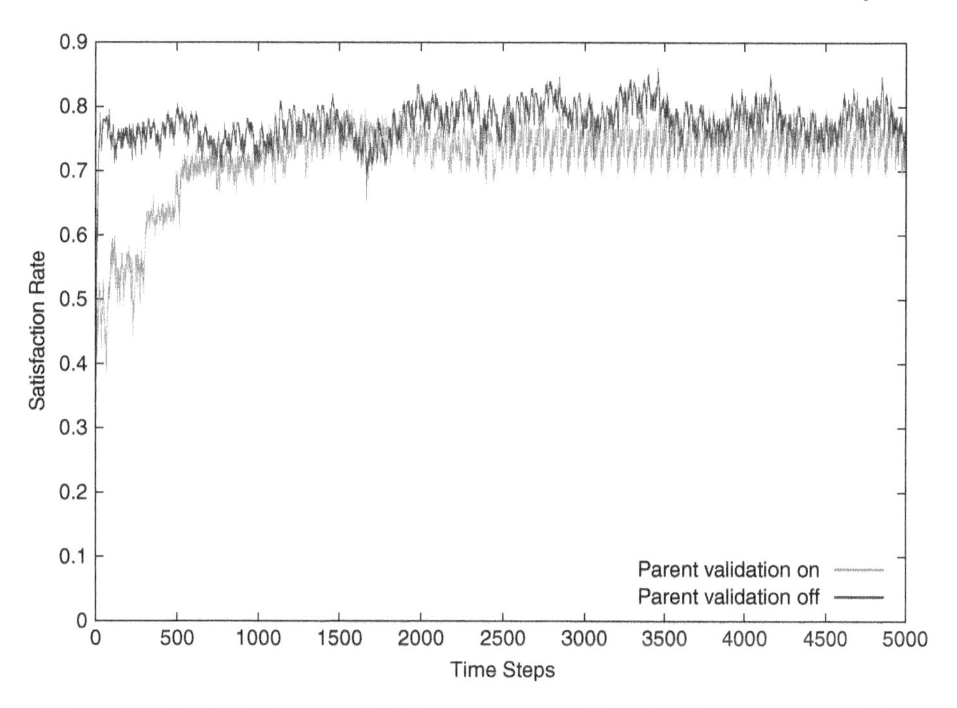

Figure 3.30 Satisfaction rate for 25 agent groups with and without parent validation

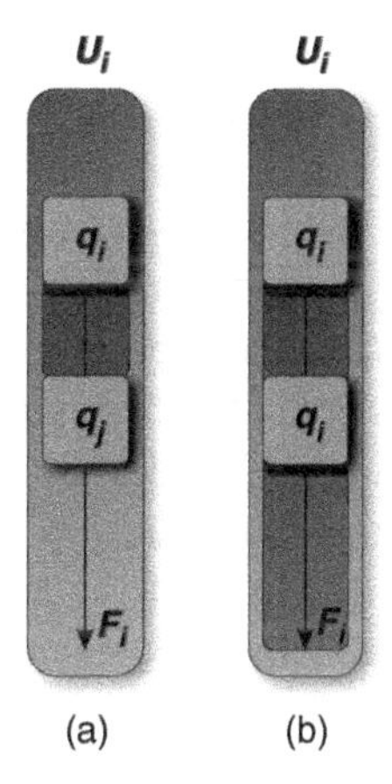

Figure 3.31 The floating improvement—adjusting the action field

14 SUMMARY

This chapter introduced intelligent control mechanisms for manufacturing and material handling systems. First, a historical overview has shown how the technological evolution and higher demands for consumers have led to new challenges for the manufacturers.

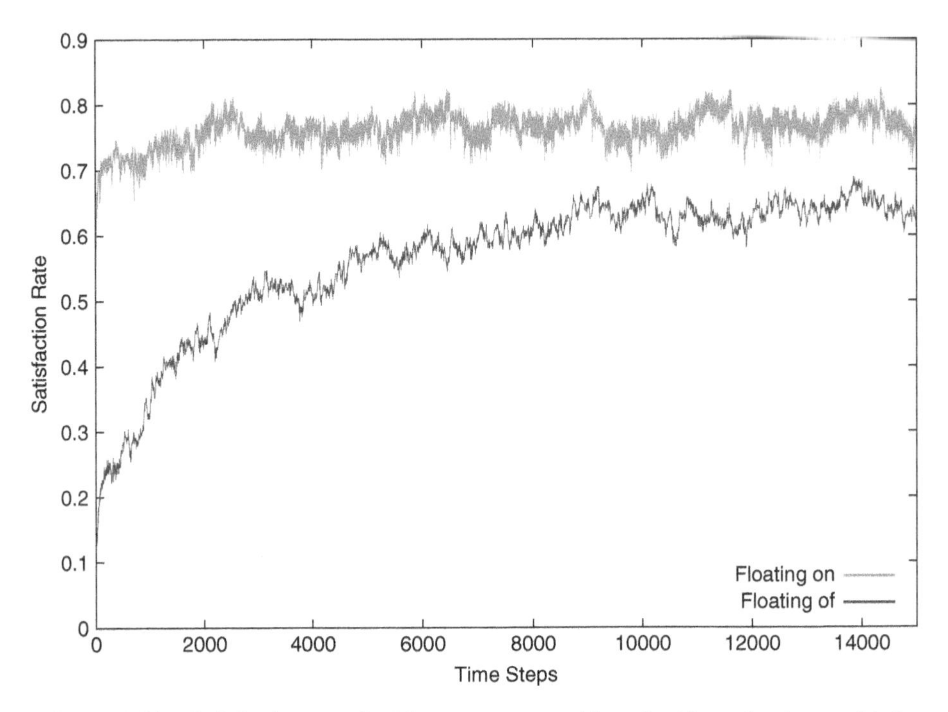

Figure 3.32 Satisfaction rate for 25 agent groups with and without floating enabled

A conceptual introduction to agent technologies was given, which summaries some of the important characteristics about both the single agent and when they collectively become a multiagent system.

Architectures, which describe the internal structure of agent, were discussion in the perspective of different agent types. Reactive agents directly respond to perceived inputs, whereas deliberative agents proactively pursue their own goals.

Also, the principles for communication and collaboration for agents were presented in the chapter. Coordination and negotiation are the dominating approaches for manufacturing and material handling systems, but in most cases a control system will use a hybrid of the two approaches. The means of collaboration are strongly influenced by the organization of the agents. This chapter introduced a number of structures to facilitate this, such as a hierarchical or team-based organization.

Finally, the chapter concluded with two practical examples of different agent-based manufacturing and material handling systems that were very different in their approach for using agent technologies: a baggage handling system composed of highly deliberative agent negotiating all the actions in the system; and a control system for transferring items between baths in a chemical process, where agents were simple reactive agents that coordinated their actions to reach a solid plan for the global system.

REFERENCES

Asai, K., S. Takashima, and P. R. Edwards. 1994. *Manufacturing automation systems and CIM factories*. London: Chapman and Hall.

Brennan, R. W., and D. H Norrie. 2001. Agents, holons and function blocks: Distributed intelligent control in manufacturing. *Journal of Applied Systems Studies Special Issues on Industrial Applications of Multi-Agent and Holonic Systems* 2(1): 1–19.

Brennan, R. W., and D. H. Norrie. 2003. From FMS to HMS. In *Agent-based manufacturing—Advances in the holonic approach*, ed. S. M. Deen. Berlin: Springer, pp. 31–49.

Brooks, R. A. 1986. A robust layered control system for a mobile robot. *IEEE Journal of Robotics and Automation* 2 (1): 14–23.

Brückner, S., J. Wyns, P. Peeters, and M. Kollingbaum. 1998. Designing agents for manufacturing control. *Proceedings of the 2nd Artificial Intelligence and Manufacturing Research Planning Workshop*, August: 40–46.

Bussmann, S. 1998. An Agent-oriented Architecture for Holonic Manufacturing Control. *Proceedings of First Workshop on Intelligent Manufacturing Systems Europe*, pp. 1–12, Lausanna, Switzerland 1998, EPFL.

Bussmann, S., N. R. Jennings, and M. Wooldridge. 2004. *Multiagent systems for manufacturing control—A design methodology*. Berlin: Springer.

Bussmann, S., and K. Schild. 2001. An agent-based approach to the control of flexible production systems. *Proceedings of 8th IEEE Int. Conf. on Emerging Technologies and Factory Automation (ETFA 2001)*, pp. 169–174, Antibes Juan-les-pins, France.

Christensen, J. H. 1994. HMS: Initial architecture and standard directions. *Proceedings of the 1st European Conference on Holonic Manufacturing Systems*, pp. 1–20, HMS Consortium, Hannover, Germany.

Colombo, A.W. 2006. An agent-based intelligent control platform for industrial holonic manufacturing systems. *IEEE Transactions on Industrial Electronics* 53 (1), (February).

Davis, S. 1989. Mass customizing. *Planning Review* 17 (2): 16–21.

Decker, K. 1995. *Environment centered analysis and design of coordination mechanisms*, Ph.D. thesis, Department of Computer Science, University of Massachusetts, May 1995.

Decker, K. 1996. *TAEMS: A framework for environment centered analysis and design of coordination mechanisms*. New York: Wiley Inter-Science, Ch. 16, pp. 429–448.

Decker, K. and J. Li. 1998. Coordinated hospital patient scheduling. *Proceedings of the Third International Conference on Multi-Agent Systems (ICMAS98)*, pp. 104–111, IEEE Computer Society, Washington, DC.

Decker, K., and J. Li. 2000. Coordinating mutually exclusive resources using GPGP. *Autonomous Agents and Multi-Agent Systems* 3 (2): 133–157.

Demazeau, Y. 1991. Coordination patterns in multi-agent worlds: Application to robotics and computer vision. *IEEE Colloqium on Intelligent Agents*. IEEE, London. February 1991.

Demazeau, Y. 1995. From interactions to collective behaviour in agent-based systems. *Proceedings of the First European Conference on Cognitive Science*, pp. 117–132. Saint Malo, France.

Di Caro, G., and M. Dorigo. 1997. *AntNet: A mobile agents approach to adaptive routing*. Université Libre de Bruxelles, IRIDIA/97-12, Belgium.

Dijkstra, E. W. 1959. A note to two problems in connexion with graphs. *Numerische Mathematik* 1: 269–271.

Donaldson, A. J. M. 2002. *A case narrative of the project problems with the Denver Airport Baggage Handling System (DABHS).* Middlesex University, School of Computing Science, TR 2002–01, January.

Erman, L. D., and V. R. Lesser. 1975. A multi-level organization for problem solving using many, diverse, cooperating sources of knowledge. Marts. *Proceedings of the Fourth International Joint Conference on Artificial Intelligence*, pp. 483–490.

Ferber, J., O. Gutknecht, and F. Michel. 2003. From agents to organizations: An organizational view of multi-agent systems. *Proceedings of AOSE'03*, Springer LNCS, vol. 2935, pp. 214–230.

FIPA. 2002a. *FIPA abstract architecture specification*. Foundation for Intelligent Physical Agents, SC00001L (December).

FIPA. 2002b. *FIPA communicative act library specification*. Foundation for Intelligent Physical Agents, SC00037J (December).

Flake, C., C. Geiger, G. Lehrenfeld, W. Mueller, and V. Paelke. 1999. Agent-based modeling for holonic manufacturing systems with fuzzy control, *Proceedings of 18th International Conference of the North American Fuzzy Information Processing Society (NAFIPS'99)*, New York, pp. 273–277.

Finin, T., J. Weber, G. Wiederhold, M. Genesereth, R. Fritzson, D. Mckay, J. Mcguire, R. Pelavin, S. Shapiro, and C. Beck. 1993. *Specification of the KQML Agent-Communication Language*. The DARPA Knowledge Sharing Initiative External Interfaces Working Group 1993.

Giret, A. and V. Botti. 2005. Analysis and design of holonic manufacturing systems, *Proceedings of 18th International Conference on Production Research (ICPR2005)*.

Gufflet, Y., and Y. Demazeau. 2004. Applying the PACO paradigm to a three-dimensional artistic creation. *Proceedings of 5th International Workshop on Agent-Based Simulation (ABS'04)*, pp. 121–126, May 2005, Lisbon, Portugal.

Hewitt, C. 1977. Viewing control structure as patterns of passing messages. *Journal Artificial Intelligence*, 8 (June).

Jenning, N. R. 1996. Coordination techniques for distributed artificial intelligence. In *Foundations of Distributed Artificial Intelligence*, ed. G.M.P. O'Hara and N. R. Jennings. New York: Wiley, pp. 187–210.

Koestler, A. 1967. *The ghost in the machine*. London: Penguin.

Kragh, A. 1990. *Kø-Netværksmodeller til Analyse af FMS Anlæg*. Ph.D. thesis, Informatics and Mathematical Modelling, Technical University of Denmark.

Maionea, G. and D. Naso 1996. A soft computing approach for task contracting in multi-agent manufacturing control. *Computers in Industry* 52(3): 199–219.

Malone, T.W. and K. Crowston 1994. The interdisciplinary study of coordination. *ACM Computing Surveys* 26: 87–119.

Malone, T. W., K. Crowston, J. Lee, B. Pentland, C. Dellarocas, G. Wyner, J. Quimby, C. Osborne, A. Bernstein, G. Herman, M. Klein, and E. O'Donnel. 1999. Tools for inventing organizations: Towards a handbook of organizational processes. *Management Science* 45(3): 425–443.

Mařik, V. and M. Pěchouček 2001. Holons and agents. Recent developments and mutual impacts. *Proceedings of the Twelfth International Workshop on Database and Expert Systems Applications*, pp. 605-607, IEEE Computer Society.

Mařik, V., M. Pěchouček, P. Vrba, and V. Hrdonka. 2003. FIPA standards and holonic manufacturing. In *Agent-Based Manufacturing—Advances in the Holonic Approach*, ed. S. M. Deen. Berlin: Springer, pp. 31–49.

Negnevitsky, M. 2005. *Artificial intelligence—A guide to intelligent systems*, 2nd ed., Reading, MA: Addison Wesley.

Parunak, H. V. D. 1987. Manufacturing experience with the contract net. In *Distributed Artificial Intelligence*, ed. M. N. Huhns. London: Pitman, pp. 285–310.

Parunak, H. V. D. 1995. *Autonomous agent architectures: A non-technical introduction*. Industrial Technology Institute, ERIM.

Parunak, H. V. D. 1996. Applications of distributed artificial intelligence in industry. In *Foundations of Distributed Artificial Intelligence*, ed. G. O'Hara and N. Jennings. New York: Wiley Interscience, pp. 139–163.

Parunak, H. V. D. 1997. Go to the Ant: Engineering principles from natural multi-agent systems. *Annals of Operation Research* 75: 69–101.

Parunak, H. V. D. 1999. Industrial and practical applications of DAI. In *Multiagent systems—A modern approach to distributed artificial intelligence*, ed. G. Weiss. Cambridge, MA: MIT Press, pp. 377–421.

Pruitt, D. G. 1981. *Negotiation Behaviour*. New York: Academic Press.

Rembold, U., and B. O. Nnaji, 1993. *Computer integrated manufacturing and engineering*. Boston, MA: Addison-Wesley 1993.

Reynolds, C. W. 1987. Flocks, herds, and schools: A distributed behavioral model. In *Computer Graphics (SIGGRAPH '87 Conference Proceedings)*, 21: 25–34.

Sandell, N., P. Varaiya, M. Athans, and M. Safonov. 1978. Survey of decentralized control methods for large scale systems. *IEEE Transactions on Automatic Control* AC-23 (2) (April): 108–128.

Scheer, August-William. 1992. *Architecture of integrated information systems: Foundations of enterprise modelling*. New York: Springer-Verlag.

Schoonderwoerd, R., O. Holland, J. Bruten, and L. Rothkrantz. 1996. Ant-based load balancing in telecommunication networks. *Adaptive Behaviour* 5: 169–207.

Searle, J. R. 1969. *Speech acts*. London: Cambridge University Press.

Shen, W., and D. Norrie. 1999. Agent-based systems for intelligent manufacturing: A state-of-the-art survey. *International Journal of Knowledge and Information Systems* 1 (2): 129–156.

da Silva, J. L. T. and Y. Demazeau. 2002. Vowels co-ordination model. *AAMAS '02: Proceedings of the First International Joint Conference on Autonomous Agents and Multiagent Systems*, New York: ACM Press, pp. 1129–1136.

Sipper, D., and R. Bulfin. 1997. *Production: Planning, control and integration*. New York: McGraw-Hill College.

Smith R.G. 1979. A framework for distributed problem solving. *Proceedings of the 6th International Joint Conference of Artificial Intelligence*, pp. 836–841.

Swamidass, P. M. 1988. *Manufacturing flexibility*, Monograph No. 2, Operations Management Association, Texas.

Taylor, F. W. 1911. *The principles of scientific management*. New York: Harpers & Brothers Publishers.

Valckenaers, P., P. V. Hadeli, M. Kollingbaum, H. van Brussel, and O. Bochmann. 2002. Stigmergy in holonic manufacturing systems. *Integrated Computer-Aided Engineering* 9 (3): 281–289.

Williamson, D. 1967. System 24—A new concept of manufacture, *Proceedings of the 8th International Machine Tool and Design Conference*, pp. 327–376, University of Manchester, September, Pergamon Press, Oxford.

Wooldridge, M., and N. Jennings. 1995. Intelligent agents: Theory and practice. *Knowledge Engineering Review* 10 (2): 115–152.

Wooldridge, M. 1999. Intelligent agents. In *Multi-agent systems*, ed. G. Weiss, Cambridge, MA: MIT Press, pp. 27–77.

Wooldridge, M. 2002. *An introduction to multiagent systems*. New York: John Wiley.

Zambonelli, F., N. Jennings, and M. Wooldridge. 2000. Organisational abstractions for the analysis and design of multi-agent systems. *In 1st International Workshop on Agent-Oriented Software Engineering*. Limerick, Ireland.

CHAPTER 4

INCORPORATING ENVIRONMENTAL CONCERNS IN SUPPLY CHAIN OPTIMIZATION

Maria E. Mayorga
Clemson University
Clemson, South Carolina

Ravi Subramanian
Georgia Institute of Technology
Atlanta, Georgia

1 INTRODUCTION

Increasing regulatory and market pressures during the past decade have fundamentally impacted supply chain decision making, starting from raw material sourcing through processing, use, and postuse—including the logistical activities in between. Linear supply chain models with unidirectional flows of materials from the upstream to the downstream links have made way for closed-loop models that necessarily involve return flows downstream to upstream (see Figure 4.1). Often, these new supply chain models involve new parameters, decision variables, constraints, and potentially conflicting and multiple objectives, translating into a need for innovative optimization methods. In this chapter, we describe how conventional supply chain optimization models have to be recast and solved differently to accommodate legislative, economic, and social pressures related to the life-cycle environmental impacts of products.

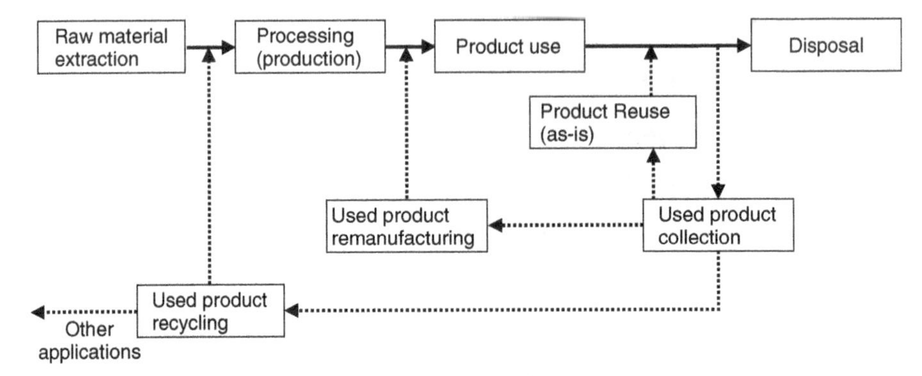

Figure 4.1 Forward and reverse flows

Supply chain optimization (SCO) refers to either the collective or individual optimization of one or more objectives by supply chain link(s) subject to various constraints, either voluntary or mandatory. With the changing landscape of dimensions along which business performance is evaluated, it is no longer obvious that profit maximization should be the sole objective in decision making. Likewise, the set of constraints to be applied extends beyond the obvious capacity and sales limits to both regulatory and voluntary impositions that echo the intricate trade-off between economic and environmental interests.

Rather than ambitiously prescribe an all-encompassing optimization model detailing every parameter, decision, and constraint that an SCO effort may incorporate, our focus in this chapter is to highlight the pertinent changes to SCO necessitated by environmental considerations so that these changes can be appropriately recognized and incorporated by supply chain managers in their decision making.

We organize this chapter around three factors—legislative, economic, and social—that have introduced environment-related complexities into SCO decisions. For each of the factors, we describe how the accompanying complexities can be characterized within SCO models in the form of parameters, objectives, or constraints. Section 2 discusses how SCO is affected by legislative instruments, such as those that prescribe design and emissions standards or impose requirements on end-of-life products. In Section 3, we discuss economic factors such as scarcity of resources, low-cost competitors, and competitive advantage arising from environmental efforts. In Section 4, we discuss social factors such as increased consumer sensitivity to the life-cycle environmental impacts of products, and stakeholder demands for voluntary environmental efforts that go beyond compliance. In Section 5, we recommend nontraditional optimization methods that are capable of accommodating the SCO refinements we propose.

2 LEGISLATIVE FACTORS

In this section, we briefly describe how environmental legislation has evolved and reflect on the accompanying changes to supply chain decision making. Environmental legislation has recently moved toward goal-oriented and market-based policies that require the beneficiaries of goods and services to incur the externalities associated with production, distribution, and consumption. In particular, regulatory instruments increasingly have two goals:

1. Address the ultimate goal of eliminating or reducing discharges into waste streams (air/water/land).
2. Create a "market price" for such discharges by requiring permits for discharges.

We classify recent legislative instruments into two categories—extended producer responsibility (EPR) and cap and trade. Although these instruments are both punitive in nature, legislation occasionally serves to encourage environmentally friendly products or processes. Tax credits for energy-efficient equipment are one example (Internal Revenue Service 2006). The SCO effort should allow for the consideration of environmentally friendly options that could become viable.

2.1 Extended Producer Responsibility (EPR)

EPR is based on the philosophy that the *beneficiaries* of products—namely, producers and customers—should be held responsible for the environmental impacts of products, rather than local governments (Lindhquist 1992). Legislative instruments specifying design standards and product take-back broadly constitute EPR. Within these two broad categories, the specific allocations of physical and financial responsibilities vary with the particular legislative implementation. We reflect on the changes that design standards and product take-back entail to SCO.

2.1.1 Design Standards

Design standards aim to *prevent* discharges into waste streams as opposed to having to mitigate them. Regulatory instruments such as the RoHS (Restriction of Hazardous Substances) directive in the European Union (EU) serve to eliminate toxic materials at-source rather than deal with these materials at end-of-life. The effect of such a regulatory tactic on SCO is the imposition of related constraints or the elimination of options such as material-process combinations (Stuart et al. 1999).

2.1.2 Product Take-Back

Similarly, product take-back focuses on the ultimate goal of preventing end-of-life products from entering into waste discharge streams. Product take-back legislation has taken various forms—from the WEEE (Waste Electrical and Electronic Equipment) directive in the EU that requires manufacturers to take back end-of-life products, to the state of California's electronic waste recycling act that charges customers a fee for end-of-life product recovery at the point of sale. Such regulatory moves and the accompanying parameters and constraints have forced supply chain managers to rethink traditional business models. In particular, the traditional supply chain optimization problem is affected by product take-back in the following ways:

- *Product mix decisions are now affected by the costs involved in recovering end-of-life products.* Some products are inherently more viable to recover than others, and might therefore become more attractive to manufacture than before.
- *Pricing decisions are affected by the fees faced by consumers and the costs of product recovery.* It is likely that optimal production quantities under product take-back are lower than those in absence of product take-back, due to increased prices.
- *Product take-back legislation has prompted businesses to evaluate alternatives to selling, such as leasing or servicizing.* By selling a service rather than the product *per se*, manufacturers have better control over the quality and recovery of end-of-life products. Manufacturers, facing the take-back of products at end-of-lease, have incentives to design their products to facilitate remanufacturing or reuse. New business models such as *servicizing* radically change the optimization problem. Instead of simply having to add new constraints or modify cost parameters in an existing problem, the structure of the optimization problem itself can change significantly (due to the need to relate variables such as product design choices, customer behavior, lease durations, and the likelihood or extent of product recovery).

2.2 Cap and Trade

Market-based programs such as emissions trading have emerged as the legislative instruments of choice to limit levels of pollutants into discharge streams. Under a permit trading program, a firm must account for each unit of pollution with a valid emissions permit. Unused permits can be banked for future use and represent a valuable, relatively liquid asset. The most successful program to date has been the U.S. Acid Rain Program to control sulfur dioxide (SO_2) emissions. Such programs offer firms the flexibility to choose amongst a variety of compliance methods and eliminate the regulatory and administrative costs typical of "command and control" instruments such as taxes. The European Union has also initiated a greenhouse gas emissions trading scheme (ETS) for controlling carbon dioxide (CO_2) emissions. With the Kyoto protocol coming into force, transnational cap and trade programs for carbon dioxide have been posited as ways to achieve targeted reductions. One such program that has been contemplated is the establishment of CO_2 emissions caps across point and distributed sources that constitute various steps in the value addition process from raw material suppliers, through manufacturers and logistics providers, to customers. Thus, firms operating in supply chains must factor new elements (decision variables, constraints, and parameters) into SCO models. We provide three examples below.

1. *Decision variables* of at-source and end-of-pipe compliance levers, number of permits to buy, sell, or bank from year to year.
2. *Constraint* on emissions; emissions cannot exceed the number of permits on hand. The shadow price corresponding to this constraint would be the value that the firm places on an incremental permit.
3. Stochastic net present value (NPV) of a banked permit, since future market value is influenced by a host of factors including compliance activity in the industry and uncertainty in future regulation.

2.3 Uncertainty in the Evolution of Environmental Legislation

In the authors' experience working with managers charged with environmental decision making in the automotive and electronics industries, uncertainty in future legislation wreaks havoc on current decision making. No longer can managers rely on static or myopic optimization approaches that are incapable of accommodating uncertainties in the optimization problem parameters (such as emissions limits, costs of compliance, and penalties for noncompliance). The supply chain optimization problem is affected by legislative uncertainty in two ways:

1. Although it is generally believed that costs of compliance (through at-source or end-of-pipe efforts) are convex increasing in the pollution outcome, future costs of compliance are difficult to predict and can materially influence current decisions. Robust or stochastic methods of optimization are therefore required.
2. Likewise, the constraints likely to be faced in the future are unknown in identity (i.e., which pollutants) and magnitude (i.e., what limits). For example, the regulatory requirement of lead-free solder in the EU (through the RoHS directive) resulted in a sudden change in supply chain decisions for most businesses dealing with electronic and electrical equipment.

3 ECONOMIC FACTORS

Economic factors such as resource scarcity, competition from low-cost producers, and the evolution of so-called *green* customer segments have recently come to the forefront and have prompted significant changes to the way SCO is to be approached. In certain instances, economically viable approaches such as remanufacturing have had positive environmental effects, such as reduced energy consumption and reduced waste. However, environmentally friendly endeavors can often be expensive relative to their economic benefits. The relationship between the costs of environmental efforts and their economic benefits has been debated at length in the literature (Walley and Whitehead 1994, Porter and van der Linde 1995). What is clear, though, is the fact that global supply chains are currently having to contend with varying intensities of economic factors. In this section, we describe how resource scarcity, competition, and the presence of "green" customer segments, influence SCO.

3.1 Resource Scarcity

Quoting from the 2008 Barclays Equity Gilt Study (Barclays Capital 2008),

> ... resource scarcity is the single most important social, political and economic factor of our era and will remain so for the foreseeable future.

Although regulation (such as product take-back) is a powerful impetus for resource conservation, resource scarcity and the accompanying price increases prompt supply chain managers to revisit traditional, purely forward models of material flows. SCO should include the evaluation of alternatives such as reuse, remanufacturing, or recycling to offset the economic disadvantages of resource scarcity or increased prices. Additionally, resource scarcity allows for the consideration of either previously eliminated or new material-processing combinations. A holistic SCO effort should plan and optimize not only the forward flows of materials but also the reverse flows. In particular, SCO is affected by resource scarcity in several ways, four of which we outline here:

1. Cost parameters and availabilities (i.e., constraint limits or RHSs) associated with various materials/components have to be dynamically updated.
2. Alternative material-processing combinations must be introduced into the optimization problem since they may become attractive under certain scenarios.
3. Planning horizons must be carefully chosen due to the inherent dynamism of commodity markets. In addition, robust methods of optimization must be chosen to accommodate such dynamism. For example, Realff et al. (2004) develop a robust mixed-integer program to support the design of an appropriate reverse production infrastructure, using the carpet industry as context. Given significant infrastructural resource commitments, their approach takes into account uncertainty in both the volumes of collected product and the prices of recovered materials.
4. Considerations must be made to allow for the assessment of the viability of product recovery. Costs of product recovery include reverse logistics costs and the costs of processing returned products. If product recovery shows promise, SCO must be expanded to jointly optimize forward and reverse supply chain activities. Many researchers have highlighted the importance of treating forward and reverse supply chain activities in

conjunction. Toktay et al. (2000) demonstrate the value of new component sourcing policies that take into account future product returns; in a later study Toktay et al. (2003) describe how hybrid manufacturing or remanufacturing production and inventory decisions can be made optimally; Debo et al. (2006), Ferrer and Swaminathan (2006), and Ferguson and Toktay (2006) discuss why and how product pricing for new and remanufactured products should be undertaken jointly; Fleischmann et al. (2001) discuss the design of the distribution network, incorporating considerations for eventual product collection; and Savaşkan et al. (2004) analyze changes to the manufacturer–retailer relationship that arise with the addition of product collection.

3.2 Competition from Low-Cost Producers

International boundaries have become relatively seamless during the past decade. As a result, domestic producers have to contend with low-cost competitors from abroad. Low-cost competitors typically have access to inexpensive labor and are subject to less stringent regulatory limitations in their respective countries. The accompanying competitive pressures necessitate changes to SCO similar to those summarized for resource scarcity in section 3.1. We list four such changes below.

1. *Considerations for alternative or new material-process combinations.* For example, successful manufacturing companies in the United Kingdom have altered their methods and procedures in order to compete with low-cost foreign imports that have saturated the UK market. Furniture makers who have implemented lean manufacturing principles have been able to minimize waste and reduce costs (Burbidge 2008).
2. *Considerations for reverse material flows.* This relates to evaluating the viability of planning for and recovering value from products during or at the end of their economic lives through reuse, remanufacture, or recycling. From the authors' experience, remanufacturing an automotive component can be 60 to 80 percent less expensive than manufacturing anew.
3. *Appropriate incorporation of intangible competitive priorities, such as the overall environmental appeal of a product.* This provides an interesting and untraditional flavor to SCO in that it entails the incorporation of subjective elements such as customer utilities that can help tilt purchasing decisions in favor of more expensive, but environmentally benign, products. Several methods have been posited in the academic literature for incorporating such customer utilities into SCO models (Chen 2001). In Section 3.3, we further discuss the need to address "green" customer segments.
4. *The presence of low-cost competition and whether and to what extent to lobby for the enforcement of regulatory standards (such as standards*

related to the environmental impacts of products). For example, *green barriers*—such as the EU's framework directive on eco-design requirements for energy-using products—serve to protect industries in developed countries from low-cost overseas competition (ChinaDaily.com.cn 2007). The likelihood of lobbying success depends on factors such as firm size and the extent of cooperation within the industry. Thus, the SCO approach must be capable of treating the risks of lobbying efforts being unsuccessful and must be able to arrive at an optimal sequence of actions (e.g., if lobbying is successful, implement decision A; if unsuccessful, implement decision B). The Mihocko, Inc. business case by Lovejoy and Cummings (1993) involves the incorporation of lobbying into decision making (although in the context of lobbying *against* emissions limits).

3.3 The Green Segment

As early as 1990, a poll by the Roper Center for Public Opinion Research found that 22 percent of U.S. respondents sought out green products and were willing to pay a premium for them (Organization 1990). In a similar 2008 poll by Harris Interactive, 47 percent of U.S. respondents indicated that they would be willing to pay more for environmentally friendly products. Moreover, these respondents indicated a willingness to spend 17 to 19 percent more for green products (Newswire 2008). Anecdotal evidence too supports the growing size of such "green" customer segments. For example, the emerging "Natural Lifestyles" market segment Lifestyles of Health and Sustainability (LOHAS), which includes home furnishings, cleaning supplies, energy-efficient lighting, apparel, and philanthropy, had an estimated U.S. market size of $10.6 billion in 2005. Given their sheer size, green customer segments cannot be ignored and must instead be consciously recognized in SCO.

1. The growing presence of such segments necessitates the incorporation of additional parameters in SCO models. Optimal product line design and market segmentation decisions should take into account consumer valuation of both environmental as well as traditional attributes (such as functionality). Chen (2001) considers the problem of designing a product line in which consumers value specific environmental attributes of the product, such as recyclability. Such a problem differs from the traditional product line design problem in several ways. Consumer utilities have to be modeled innovatively, since a segment of customers also receives intangible benefits from consuming green products. However, environmental attributes of the product could conflict with its traditional quality attributes (e.g., with diesel engines, greater material thicknesses facilitate re-surfacing during remanufacturing steps, but at the expense of functional quality as measured by the power-to-weight ratio).

2. New decision variables, such as marketing effort for communicating the environmentally positive elements of products, must enter the objective function.
3. The SCO approach must be capable of capturing the dynamism inherent in customer segments. For instance, the assortment (consisting of a mix of standard and green products) must change with changes in customer utilities for green products.

4 SOCIAL FACTORS

Coupled with the recent growth of green consumer segments described in Section 3.3 is an increased societal awareness of the life-cycle impacts of products spanning various steps of value addition. Various stakeholders such as customers, employees, shareholders, communities, lobbyists, and local organizations increasingly demand more than just regulatory compliance and instead expect businesses to proactively undertake environmental initiatives despite the expense. Increased media attention to environmental issues has further reinforced these expectations. Proactive environmental initiatives may not have immediate pay-offs but could serve to minimize a firm's liability risks (Snir 2001). Avoidance of such risk could also be treated as an economic factor.

A major challenge in responding to social pressures is identifying the constituencies that a business must respond to. In other words, it is unclear whom the firm should negotiate with or whom the firm should accommodate. For example, while clean air is good for communities in the vicinity of manufacturing locations, shareholders might be less demanding with respect to air quality if related efforts erode stock value. Consumers themselves are heterogeneous in their attention to the different environmental dimensions of products. Gallup's "Health of the Planet Survey" conducted in 1992 across 22 countries (Institute 1992) showed that consumers in industrialized countries tended to avoid environmentally harmful products; the percentages of respondents indicating such avoidance ranged from 40 percent in Japan to 81 percent in Germany; the United States came in at 57 percent. However, the same survey also showed that Germans were much more concerned about the loss of rainforests than Americans (80 percent vs. 63 percent), while respondents from both the countries were equally concerned about air pollution (60 percent and 61 percent).

In absence of directives limiting environmental impacts, third-party certifications (by consumer groups, the government, or other independent organizations) and advertising emerge as alternatives for businesses to signal their positive environmental intent. Examples of third-party certifications include "Green Seal" and the U.S. EPA's "Energy Star." The not-for-profit organization Green Seal considers environmental impacts across the firm's supply chain in certifying its product(s). For example, one of the requirements in the "Printing and Writing Paper" category is that the product must meet either specified recycled content or production process requirements (Seal 1999). The U.S. EPA states that one of the reasons for the inception of its voluntary labeling program, Energy Star, is that it was " ... designed to identify and promote energy-efficient products to reduce greenhouse gas emissions" (Star 1999). Although the program is voluntary, institutional as well individual buyers often prefer or demand that the products they purchase be Energy Star qualified in order to be assured of cost savings from reduced energy consumption, or to take advantage of tax incentives for energy-efficient purchases (e.g., Internal Revenue Service 2006). SCO should be expanded to allow for choosing amongst a myriad of certifications, given their different emphases and the associated costs of certification. Advertising campaigns can be quite costly as well. For example, Siemens AG rolled out a $145 million "Siemens Answers" campaign in late 2007 to promote the company's focus on health and the environment (Maddox 2007). Below, we summarize the ways by which the aforementioned social factors have brought about changes to SCO.

1. In the midst of societal influences, the SCO endeavor must allow for multiple, prioritized objectives. Priorities (or weights) for possibly conflicting objectives must be established after a careful analysis of what would have the greatest positive impact on the long-term economic and environmental sustainability of the business. For example, the goal programming approach (a branch of multiobjective optimization, which will be further discussed in section 5) can be used to suitably accommodate both profit maximization as well as emissions minimization objectives.
2. Related to the prioritization of objectives, new decision variables must be factored into SCO models, such as the particular environmental certifications to pursue and the extent of advertising effort to undertake. Mixed-integer methods could be applied to determine the most viable certifications. Additionally, relationships such as those between advertising effort and the resulting economic benefit (or risk avoidance) are difficult to estimate. They require robust methods of optimization.
3. The time horizons for SCO have to be altered since a firm's beyond-compliance environmental efforts typically take a longer time to be recognized and valued by the market than immediate economic gains. Because elements such as risk avoidance are difficult to precisely value, robust methods of optimization must be used to allow for uncertainties across longer time horizons.
4. SCO exercises must be expanded in scope both vertically (across links within a supply chain) as well as horizontally (across firms in different supply chains). For example, Cruz (2008) presents an integrated model of a supply chain network consisting of a manufacturer, a retailer, and consumers. This network aims to achieve the objectives of maximizing profit, minimizing emissions, and minimizing risk. Costs of socially

responsible activities (such as reduced packaging and joint recycling efforts) are a function of the extent of shared responsibility within the supply chain. Vachon and Klassen (2006b) study the operational performance of a supply chain operating under a *green project partnership*, involving both upstream and downstream efforts to prevent pollution. Using data from the packaging industry, they find that downstream (i.e., closer to the customer) green project partnerships are positively linked to quality, flexibility, and environmental performance, while upstream partnerships are associated with better delivery performance. A later study by Vachon and Klassen (2006a) finds that collaboration in the supply chain plays an even more important role as corporations attempt to gravitate toward environmental sustainability. In line with their previous study, they find that a firm's collaboration with its suppliers is closely linked with process-based performance, while collaboration with customers is associated with product-based performance. With respect to horizontal collaboration, decisions related to the endorsement of specific certifications should be collaboratively conducted across possibly competing firms, given the ambiguities associated with the myriad of standards and certifications present in industries and the noisy market signals that accompany them.

5 APPROACHES TO OPTIMIZATION

In general terms, an optimization problem involves a decision (or variable) space and a criteria (or objective) space. The appropriateness of approaches used to solve an optimization problem depends on the characteristics of these spaces. The preceding sections discussed how the three factors—legislative, economic, and social—entail changes to the decision space and/or the criteria space of an SCO problem. For example, the traditional SCO objective of solely maximizing profits fails to incorporate intangible benefits derived from environmental efforts. Traditional variables (such as product-process combinations) may have to factor in uncertainties in the evolution of environmental regulation. Thus, recent trends have created a need for the use of alternative approaches to formulating SCO problems. In this section, we provide recommendations of optimization methods that are capable of accommodating the SCO complexities discussed in the preceding sections.

5.1 Nonlinear Programming (NLP)

Nonlinear dependencies, such as between decisions and outcomes of interest, are inevitable in practice. For example, emissions abatement costs are typically convex increasing in the extent of abatement desired. Nonlinear dependencies also often exist among the decision variables themselves. For example, production decisions related to the mix between new and remanufactured products are linked nontrivially to the product design decision of product remanufacturability. Traditional, linear methods are not equipped to provide an adequate representation of such dependencies. Nonlinear methods, on the other hand, allow for more realistic representations of managerial trade-offs and the interactions among decisions and consequences. By definition, an optimization problem is nonlinear when either the objective or any of the constraints is nonlinear.

Researchers have recently demonstrated the use of nonlinear programming (NLP) methods to incorporate environmental considerations into SCO. For example, Subramanian et al. (2008) develop a nonlinear optimization model for a manufacturing firm attempting to integrate environmental considerations (such as remanufacturing and product design) with traditional operations planning considerations (such as production and inventory). The NLP approach allows them to include dependencies between new and remanufactured products such as the cannibalization of new products by remanufactured products and competition for limited production capacity. Apart from the nonlinear objective, other nonlinear elements in the optimization problem include the cost of product design and consumer demand, which is a nonlinear decreasing function of production quantities (again, decision variables). Relationships expressed in the form of nonlinear expressions can realistically and flexibly characterize a representative problem. Although the computational effort involved in solving NLPs is significant, present-day desktop computers, together with commercially available software tools, can be used to solve even complex NLPs in a reasonable amount of time, allowing for what-if analyses.

5.2 Multiobjective Optimization

Environmental considerations often give rise to multiple, possibly conflicting, objectives in SCO. For example, in the work by Subramanian et al. (2007), the firm might want to maximize profits as well as minimize emissions. A multiobjective (or multicriteria) optimization problem (MOP) involves the simultaneous optimization of two or more objectives subject to constraints. To date, the concept has been employed effectively in scenarios when problems are characterized by decisions that are conflicting in nature, which is often the case when incorporating environmental considerations. As an example, minimizing emissions is often at odds with maximizing profit. Several solution methods exist to solve multiobjective problems. This section discusses two such methods.

5.2.1 Weighted-Sum Method

One approach is to construct a single aggregate objective function (AOF), in which all of the criteria being considered are combined into a single objective. There are several ways to aggregate criteria; the weighted linear sum of objectives is often used. In this method, the decision maker specifies fractional weights for each individual objective. The AOF combines the weighted individual objectives into a single function that can be solved in the same manner as a traditional, single-objective problem. A challenge is to arrive at meaningful weights for the various objectives (Ehrgott 2005). Shue (2006) employs a MOP to optimize both generation operations and associated induced-waste reverse logistics at a nuclear facility. The AOF to be maximized aggregates the positively weighted power supply chain profits and the negatively weighted reverse chain costs. A menu of solutions is generated for varying values of weights.

5.2.2 Constraint Method

Apart from the weighted-sum approach, other techniques to solve multicriteria optimization problems also exist (e.g., ϵ-constraint method, elastic constraint method, hybrid method, etc.). Popular among these techniques is the ϵ-constraint method. Unlike the weighted-sum approach, no aggregation of criteria is necessary. Instead, one criterion is chosen as the primary criterion to be optimized and all other criteria are transformed into constraints, such as in Subramanian et al. (2008), where emissions are minimized subject to a reservation level of profit. The Greek letter ϵ denotes the RHSs (or reservation levels) of the constraints corresponding to the nonprimary objectives. For convex optimization problems (i.e., where the decision and criteria spaces are convex), the weighted-sum method is guaranteed to find efficient and weakly efficient solutions, while the ϵ-constraint method only guarantees weakly efficient solutions (Ehrgott 2005, p. 98). However, the ϵ-constraint method works for even nonconvex optimization problems.

MOPs are also well-suited for representing and solving optimization problems that span multiple firms or networks of firms and involve objectives at various levels. Sabri and Beamon (2000) develop a multiobjective optimization model for a traditional supply chain, allowing for metrics (derived not only from cost but also from other performance criteria such as customer satisfaction) to be assessed across the entire supply chain network rather that at the individual firm level. We suggest that such models can be expanded to incorporate green metrics, affording a supply-chain-wide treatment of environmental considerations.

5.3 Dynamic Models

Static (or myopic) SCO approaches are incapable of suitably accommodating the dynamism inherent in environmental factors. As mentioned in the preceding sections, factors such as legislative changes to emissions limits and the decreasing availability of certain resources render dynamism to an SCO exercise.

Several solution methods exist for complex dynamical systems. Among them are differential equations and dynamic programming. As an example, Cruz (2008) develops a dynamic framework for modeling a multilevel supply chain network. In particular, each decision maker in the network faces a multicriteria (including environmental criteria related to corporate social responsibility) decision-making problem. The dynamic system analysis aims to compute the trajectory of product flows within the network, the levels of chosen social responsibility, and product prices. The problem formulation involves differential equations that capture system dynamics embedded within an optimization problem. They propose a solution algorithm based on the Euler method. The Euler method has been applied effectively to other dynamic supply chain network problems as well (Nagurney and Dong 2002).

The rolling time-horizon approach, in which the start and the end of the planning horizon are rolled forward as time progresses, has been employed for dynamic decision making in the context of traditional supply chain decisions: optimizing a vendor-managed inventory system (Al-Ameri et al. 2008), designing flexibility contracts under variable demand (Walsh et al. 2007), and setting safety-stocks in a multistage inventory system (Boulaksil et al. 2007), to name a few. Such problems are solved using either traditional mathematical programming techniques, simulation, or heuristic methods. Environmental considerations that necessitate a rolling time-horizon approach can be accommodated in a similar fashion.

5.4 Stochastic Programming and Robust Optimization

Decisions related to the environment often involve a high level of uncertainty. The source of this uncertainty ranges from resource availability to future regulatory requirements. Deterministic optimization methods (such as linear programming) assume that objective coefficients and constraints are known with certainty. Deterministic methods may serve as good proxies if good estimates of the unknown parameters exist. By employing sensitivity analysis, the decision maker can determine the ranges of parameters over which a solution is feasible or even optimal. However, sensitivity analysis is not particularly helpful for generating solutions that are robust to parameter changes.

The stochastic programming approach directly deals with uncertainty, in that the uncertain parameters are modeled as random variables with known probability distributions. The goal, then, is to achieve the best objective value in expectation. A popular method is the stochastic linear programming model with *recourse*, in which corrective action can be taken after uncertainty is resolved. For example, a two-stage recourse model has decision variables in each of the two stages. The second-stage variables relate to recourses available after the realization of uncertainty. In stochastic optimization, the problem is expressed using probabilistic constraints—the assumption being that the distributions of problem parameters are known. However, in practice, it may be difficult to obtain or even approximate parameter distributions. In addition, stochastic optimization models are often difficult to analyze. For more on stochastic programming in the context of supply chains, see Shanthikumar et al. (2003).

Robust optimization is an alternative to deal with the drawbacks of stochastic optimization. Under robust optimization, uncertain parameters are known only to belong to an "uncertainty" set (i.e., knowledge of their specific distributions is not necessary). The focus is on generating solutions that are *robust* or immune to parameter changes. Early concepts of robust optimization, such as the one proposed by Soyster (1973) in the early 1970s, sought to construct optimization models such that the solution would be feasible over entire uncertainty sets. The drawback of such models is that solutions are too conservative in the sense that the objective is sacrificed for robustness. Recent concepts of robustness involve the consideration of worst-case scenarios. Mulvey et al. (1995) present a min–max objective approach that integrates goal programming formulations with scenario-based data descriptions for optimizing against worst-case scenarios.

5.5 Optimality Criteria

The concept of *optimality* in SCO may have to be recast in light of societal pressures (such as issues of "fairness") and due to concerns of supplier reliability leading to diversification efforts. One approach in the literature is the concept of *equitable efficiency*, which can be applied to both single and multiobjective optimization problems. Traditionally, decision makers are interested in solutions that are Pareto-optimal or efficient (i.e., no other solution yields a strictly better result). However, this optimality criterion may not be appropriate when dealing with environmental concerns or other issues that are difficult to quantify. During the past decade, an increasing interest in equity issues has resulted in new methodologies in the area of operations research; equitable efficiency has been proposed as one refinement to the concept of Pareto optimality.

To understand the concept of equitable efficiency, consider a typical optimization problem. The optimality of a solution is traditionally determined by the magnitude of the resulting objective value. In the case of equitable efficiency, the decision maker not only is interested in the value of the objective but also desires to achieve a balance or fairness among outcomes. Consider an illustrative example of a multiobjective problem, similar to that in Kostreva et al. (2004). Suppose two possible solutions exist, generating outcome vectors (6, 2, 6) and (1, 3, 1), respectively (assume that the outcome vector elements are normalized and are therefore comparable). Both solutions are Pareto-optimal (i.e., neither solution strictly dominates the other). Though both solutions have two outcome elements equal to each other, the second outcome vector is clearly better in terms of distribution of outcomes (outcomes deviate by a maximum of 2 as opposed to 4), and the solution with outcome vector (1, 3, 1) is said to equitably dominate the other solution. Thus, equitably efficient solutions are a subset of Pareto-optimal solutions.

The concept of equitable efficiency can be also be applied to MOPs in which the objectives are incomparable or even conflicting. Traditional optimality concepts are not equipped to deal with the issue of fairness, and often, various aggregations of criteria are applied in order to select suitable solutions. However, such aggregations force comparability of criteria (as in the weighted-sums method). In their paper, Kostreva et al. present criteria aggregation methods that can be used to derive equitably efficient solutions to both linear and nonlinear multiobjective problems. As methods for achieving equitable efficiency continue to be developed, they may be used in the context of environmentally conscious SCO, where issues of fairness and equitable use of resources abound.

5.6 Selecting an Appropriate Method

NLP methods should be employed when nonlinear dependencies exist among decision variables, or when problem parameters are governed by nonlinear functions. MOP methods are appropriate when the SCO exercise involves several, possibly conflicting, objectives. A point to note here is that if the various criteria are not comparable and an AOF method is chosen, then each criterion must be scaled appropriately (Feyzan and Zulal 2007). The dynamic nature of environmental factors can be accounted for by carefully choosing time-horizon lengths and modeling interdependencies among variables and time. Stochastic programming and robust optimization are designed to explicitly deal with uncertainty; however, the complexity involved can lead to tractability and computational issues. Lastly, "optimal" SCO decisions are materially impacted by the choice of optimality criteria (e.g., an "equitably efficient" solution does not necessarily result in the greatest profit or value).

In concluding this section, we note that the suggested optimization approaches are not intended to serve as an exhaustive list. As SCO problems continue to increase in complexity, decision makers may have to reach for meta-heuristic solution techniques such as genetic algorithms, tabu search, and simulated annealing (Michalewicz and Fogel 2000).

6 SUMMARY

Economywide surveys unequivocally indicate that the costs of engaging in environmental efforts—mandated as well as voluntary—are indeed significant. A recent survey by the U.S. Census Bureau (April 2008) shows that capital expenditures for pollution abatement and control in the United States totaled $5.91 billion in 2005, of which $3.88 billion was attributed to air emissions, $1.35 billion to water discharge, and $0.68 billion to solid waste. Operating costs for pollution abatement and control totalled $20.68 billion. Of this, $8.63 billion was attributed to air emissions, $6.73 billion to water discharge, and $5.32 billion to solid waste.

The significance of these costs reinforces the need for an SCO effort to appropriately incorporate the factors identified in this chapter. Evidently, environmental considerations introduce fair levels of subjectivity in SCO, such as in the estimation of parameters, in the choice of objectives to pursue and their relative priorities, and in the characterization of constraints to impose. This subjectivity—coupled with uncertainties in the evolution of environmental legislation or in the likelihood of environmental efforts translating into economic or reputational benefits—demands new approaches to SCO, such as those highlighted in section 5. A sound SCO exercise requires multiple disciplines to be involved, such as strategy (e.g., to decide on competitive priorities and objectives), marketing (e.g., to relate marketing effort to reputational benefits), operations (e.g., capacity planning for new and remanufactured products), finance (e.g., assessing market valuation of environmental efforts), and accounting (e.g., deciding ROI and pay-back period benchmarks for environmental investments) (Hoffman 2005).

In conclusion, this chapter considers how SCO can be adapted when legislative, economic, and social factors related to the environment affect the fundamental SCO problem elements—parameters, objectives, and constraints. Environmental factors will, however, continue to influence SCO, and supply chain analysts must keep pace with the impacts of these factors and must appropriately adapt modeling and analysis methods.

REFERENCES

Al-Ameri, T. A., N. Shah, and L. G. Papageorgiou. 2008. Optimization of vendor-managed inventory systems in a rolling horizon framework. *Computers and Industrial Engineering* 54(4): 1019–1047.

Barclays capital. 2008. Barclays capital launches the 2008 equity gilt study. http://www.barcap.com/sites/v/index.jsp?vgnextoid=31c46798c3118110VgnVCM2000001613410aRCRD&vgnextchannel=1c6c15cd3f4f8010VgnVCM1000002581c50aRCRD. Last accessed June 2, 2008.

Boulaksil, Y., J. C. Fransoo, and E. N. G. van Halm. 2007. Setting safety stocks in multi-stage inventory systems under rolling horizon mathematical programming models. *OR Spectrum* (May). DOI: 10.1007/s00291-007-0086-3.

Burbidge, B. 2008. Follow that car firm. *Cabinet Maker* (March 14).

Chen, C. 2001. Design for the environment: A quality-based model for green product development. *Management Science* 47(2): 250–263.

ChinaDaily.com.cn. 2007. Green barrier disguises face of protectionism. August 15. http://www.chinadaily.com.cn/bizchina/2007-08/15/context_6027990.htm.

Cruz, J. M. 2008. Dynamics of supply chain networks with corporate social responsibility through integrated environmental decision-making. *European Journal of Operational Research* 184: 1005–1031.

Debo, L. G., L. B. Toktay, and L. N. Van Wassenhove. 2006. Joint life-cycle dynamics of new and remanufactured products. *Production and Operations Management* 15(4): 498–513.

Ehrgott, M. 2005. *Multicriteria optimization*, 2nd ed. Berlin, Germany: Springer.

Energy Star. 2008. About Energy Star. http://www.energystar.gov/index.cfm?c=about.ab_index. Last accessed May 26, 2008.

Ferguson, M., and L. B. Toktay. 2006. The effect of competition on recovery strategics. *Production and Operations Management* 15(3): 351–368.

Ferrer, G., and J. M. Swaminathan. 2006. Managing new and remanufactured products. *Management Science* 52(1): 15–26.

Feyzan, A., and G. Zulal. 2007. A two-phase approach for multi-objective programming problems with fuzzy coefficients. *Information Sciences* 177: 5191–5202.

Fleischmann, M., P. Beullens, J. M. Bloemhof Ruwaard, and L. N. Van Wassenhove. 2001. The impact of product recovery on logistics network design. *Production and Operations Management* 10(2): 156–173.

George H. 1992. The health of the planet survey. Technical report, Gallup International Institute, Princeton, NJ, 1992.

Green Seal. 1999. Green seal environmental standard for printing and writing paper. November 12. http://www.greenseal.org. Last accessed May 26, 2008.

Hoffman, A. J. 2005. *Business decisions and the environment: Significance, challenges, and momentum of an emerging research field*. G. D. Brewer and P. C. Stern, eds. Washington, D.C.: The National Academies Press.

Kostreva, M. M., W. Ogryczak, and A. Wierzbicki. 2004. Equitable aggregations and multiple criteria analysis. *European Journal of Operational Research* 158(2): 362–377.

Lindhquist, T. 1992. *Extended producer responsibility as a strategy to promote cleaner products*. Sweden: Department of Industrial Environmental Economics, Lund University.

LOHAS Lifestyles of Health and Sustainability. 2008. About LOHAS—A history of the sustainable marketplace. http://www.lohas.com/about.html. Last accessed June 3, 2008.

Lovejoy, W. S., and C. Cummings. 1993. Mihocko, Inc. Business Case. Stanford, CAC Leland Stanford Junior University.

Maddox, K. 2007. An inconvenient lack of metrics; Green marketings influence proves difficult to measure. *B to B Magazine* (April 7).

Metalprices.com. 2008. Free crude oil price charts. http://www.metalprices.com/pubcharts/Public/CrudeOil_Price_Charts.asp?WeightSelect=LB&SizeSelect=M&ccs=1&cid=0. Last accessed June 2, 2008.

Michalewicz, Z., and Fogel, D. 2000. *How to solve it: Modern heuristics*, 2nd ed. Berlin: Springer.

Mulvey, J., R. Vanderbei, and S. Zenios. 1995. Robust optimization of large-scale systems. *Operations Research* 43: 264–281.

Nagurney, A., and J. Dong. 2002. *Supernetworks: Decision-making for the information age*. Cheltenham, England: Edward Elgar Publishers.

Porter, M. E., and C. van der Linde. 1995. Toward a new conception of the environment-competitiveness relationship. *Journal of Economic Perspectives* 9(4): 97–118.

PR Newswire. 2008. SCA survey conducted by Harris Interactive® shows that despite a weakened economy, U.S. consumers willing to spend green to go green. *PR Newswire* (April 21).

Realff, M. J., J. C. Ammons, and D. J. Newton. 2004. Robust reverse production system design for carpet recycling. *IIE Transactions* 36: 767–776.

Sabri, E. H., and B. M. Beamon. 2000. A multi-objective approach to simultaneous strategic and operational planning in supply chain design. *OMEGA International Journal of Management Science* 28: 581–598.

Savaşkan, R. C., S. Bhattacharya, and L. N. Van Wassenhove. 2004. Closed-loop supply chain models with product remanufacturing. *Management Science* 50(2): 239–252.

Shanthikumar, J. G., D. D. Yao, and W. H. M. Zijm, eds. 2003. *Stochastic modeling and optimization of manufacturing systems and supply chains*. International Series in Operations Research & Management Science. Boston, MA: Kluwer Academic Publishers.

Shue, J.-B. 2006. Green supply chain management, reverse logistics and nuclear power generation. *Transportation Research Part E: Logistics and Transportation Review* 44(1): 19–46.

Snir, E. M. 2001. Liability as a catalyst for product stewardship. *Production and Operations Management* 10(2): 190–206.

Soyster, A. L. 1973. Convex programming with set-inclusive constraints and applications to inexact linear programming. *Operations Research* 21: 1154–1157.

Stuart, J. A., J. C. Ammons, and L. J. Turbini. 1999. A product and process selection model with multidisciplinary environmental considerations. *Operations Research* 47(2): 221–234.

Subramanian, R., B. Talbot, and S. Gupta. 2008. An Approach to Integrating Environmental Considerations within Managerial Decision-Making. http://ssrn.com/abs_acr=1004339.

Sweeney, D. 2008. Meeting the challenge of resource scarcity; Grant Thornton Weath Protection. *Birmingham Post* (April 29), p. 4.

The Roper Organization. 1990. *The environment: Public attitudes and individual behavior*. Technical report. Study commissioned by S.C. Johnson & Son, Inc, July.

Toktay, L. B., L. M. Wein, and S. A. Zenios. 2000. Inventory management for remanufacturable products. *Management Science* 46(11): 1412–1426.

Toktay, L. B., L. M. Wein, and S. A. Zenios. 2003. A new approach for controlling a hybrid stochastic manufacturing/remanufacturing system with inventories and different leadtimes. *European Journal of Operational Research* 147: 62–71.

U.S. Department of the Treasury, Internal Revenue Service. 2006. Treasury and IRS Provide Guidance for Energy Credits for Homeowners. http://www.irs.gov/newsroom/article/0,,id=154657,00.html. Last accessed May 26, 2008.

U.S. Census Bureau. 2005. Pollution abatement costs and expenditures: 2005. http://www.census.gov/prod/2008pubs/ma200-05.pdf. April 2008.

Vachon, S., and R. D. Klassen, 2006a. Environmental management and manufacturing performance: The role of collaboration in the supply chain. *International Journal of Production Economics* 111: 299–315.

Vachon, S., and R. D. Klassen. 2006b. Green project partnership in the supply chain: The case of the package printing industry. *Journal of Cleaner Production* 14: 661–671.

Walley, N., and B. Whitehead. 1994. It's not easy being green. *Harvard Business Review* (May–June): 46–52.

Walsh, P. M., P. A. Williams, and C. Heavey. 2007. Investigation of rolling horizon flexibility contracts in a supply chain under highly variable stochastic demand. *IMA Journal of Management Mathematics* 19(2): 117–135.

CHAPTER 5

MUNICIPAL SOLID WASTE MANAGEMENT AND DISPOSAL

Shoou-Yuh Chang
DOE Samuel Massie Chair Professor
North Carolina A&T State University
Greensboro, North Carolina

1 INTRODUCTION

Human activities in using the resources on earth unavoidably generate waste. Solid waste can be defined as any waste that is solid or semi-solid that is unwanted and discarded for disposal. Municipal solid waste (MSW) consists of everyday items such as package wrappings, grass clippings, furniture, clothing, bottles, food scraps, newspapers, consumer electronics, and appliances. In general, MSW does not include industrial, hazardous, or construction and demolition waste. Despite improvements in waste reduction and recycle, MSW management remains a constant concern because the generation trend indicates that the overall tonnage we generate continues to increase, especially for developing countries.

The amount of MSW generated in United States increased 60 percent from 1980 to 2006. In 2006, U.S. residents, businesses, and institutions produced more than 246 million tons of MSW, which is approximately 4.5 pounds of waste per person per day (U.S. EPA 2007). Paper and paperboard products constitute about 34 percent of the MSW stream. This is the largest portion of MSW. In 2006, Americans generated about 85 million tons of paper products, which are nearly a threefold increase from 1960. About 52 percent of all paper and paperboard products were recovered in 2006, nearly two and a half times the percentage in 1960 (U.S. EPA 2007).

Municipal solid waste is collected by cities and counties and then potentially separated into three waste streams. These waste streams are then processed as follows. Recyclable material is separated by category (paper, glass, aluminum, etc.) and then sorted and sold to brokers or vendors. Organic wastes such as yard wastes can be composted using microorganisms to produce a humuslike substance, which can be used in gardening and landscaping applications. The remaining waste stream is often placed in a landfill or sent to incinerators. The major legislation that governs the management of solid waste is the Resources Conservation and Recovery Act (RCRA) of 1976. The intent of this legislation is to promote environmentally sound solid waste management practices to maximize resource recovery. It has been amended about every couple of years. RCRA gives the legal basis for implementation of guidelines and standards for solid waste storage, treatment, and disposal. In this legislation, the U.S. EPA separated hazardous waste from municipal solid waste so that hazardous waste is regulated under Subtitle C and solid waste is under Subtitle D. Although solid waste is regulated mostly by state and local governments, the U.S. EPA has promulgated solid waste regulations to address how disposal facilities should be designed and operated. Its primary role is in setting national goals and providing technical assistance, as well as developing educational materials.

Management of municipal solid waste is a complicated problem that must be addressed to meet public health, environmental, and economical concerns. The U.S. EPA's tiered integrated waste management strategy includes the following: (1) source reduction (or waste prevention), including reuse of products and on-site (or backyard) composting of yard trimmings; (2) recycling, including off-site (or community) composting; (3) combustion with energy recovery; and (4) disposal through landfilling or combustion without energy recovery. The rest of this chapter will discuss the following aspects of MSW: sources and compositions, management processes that includes source reduction, collection, recycling, composting, incineration and landfilling, management in other countries, planning issues and optimization models development, environmental impact and life-cycle assessment, and future trends in MSW management.

2 SOURCES AND COMPOSITION OF MUNICIPAL SOLID WASTE

Municipal solid waste (MSW) is a diverse classification that includes residential, commercial, and institutional waste. Residential facilities are single and multi-family dwellings, as well as high-rise apartments and dormitory-style housing. Residential waste typically includes paper, aluminum cans, ferrous materials, glass, wood, and yard waste. Commercial waste is produced from stores, office buildings, restaurants, and other businesses. These commercial facilities can produce waste that includes paper, cardboard, glass, metals, and food waste. Institutional waste includes waste generated from schools, hospitals, and other public offices. The waste material from these sources would resemble the waste from commercial facilities. The waste generated from residential, commercial, and institutional sites are similar, with the biggest differences being in the percentages of each type of waste (Tchobanoglous et al. 1993) (see Table 5.1).

Table 5.1 shows the municipal solid waste generated in the United States from 1960 to 2006. The amount of MSW generated in United States has increased about 285 percent from 1960 to 2006. However, most of the increases, from 88 to 238 million tons, which is about 270 percent, occurred between 1960 and 2000. Because of the environmental concerns and intensive waste reduction and recycling effort, the increase between 2000 and 2006 is about 5.5 percent, from 238 to 251 million tons. If the U.S. population increase is taken into account, most of the increase occurred between 1960 and 1990 and the per capita solid waste generation since 1990 is between 4.1 to 4.6 pounds per day. In 2006, U.S. residents, businesses, and institutions produced more than 251 million tons of MSW, which is approximately 4.5 pounds of waste per person per day (U.S. EPA 2007). Residential waste (including waste from apartment houses) is estimated to be 55 to 65 percent of the total municipal solid waste generation, while waste from schools and commercial locations, such as hospitals and businesses, amounted to 35 to 45 percent.

Table 5.2 shows the typical composition of MSW generated in the United States from 2000 to 2006 based on Table 5.1. The total MSW generation in 2006 was 251 million tons. Organic materials continue to be the largest component of MSW. Figure 5.1 illustrates the composition of MSW in 2006. Paper and paperboard products account for 33.9 percent. Yard trimmings are the second-largest component at 12.9 percent, followed by the food waste of 12.4 percent. Plastics are 11.7 percent, metals make up 7.6 percent, and glass accounts for 5.3 percent. The rest of 16.1 percent comes from wood, textiles, rubber, leather, and other miscellaneous waste. Figure 5.2 shows the trend of generation and composition of MSW from 1960 to 2006. Although paper and paperboard products have been the largest component of the MSW, the amount became steady and does not show any increase since year 2000. However, the amount of plastics continues to increase over the years, from less than 1 percent in 1960 to 11.7 percent

Table 5.1 Municipal Solid Waste Generated in the United States from 1960 to 2006 (in 1000 tons*)

Waste Component	1960	1970	1980	1990	1995	2000	2001	2002	2003	2004	2005	2006
Paper and Paperboard	29,990	44,310	55,160	72,730	81,670	87,740	82,660	84,070	83,030	87,550	85,130	85,290
Glass	6,720	12,740	15,130	13,100	12,830	12,620	12,580	12,570	12,340	12,650	12,760	13,200
Metals	10,820	13,830	15,510	16,550	15,860	18,240	18,280	18,310	18,770	18,810	18,680	19,130
Plastics	*390*	2,900	*6,830*	17,130	18,900	25,340	25,270	27,180	27,620	29,210	28,950	29,490
Rubber and Leather	1,840	2,970	4,200	5,790	6,030	6,530	6,670	6,660	6,820	6,690	6,670	6,540
Textiles	1,760	2,040	2,530	5,810	7,400	9,440	9,810	10,320	10,590	10,930	11,280	11,840
Wood	3,030	3,720	7,010	12,210	12,780	13,020	13,180	13,340	13,610	13,730	13,900	13,930
Food Waste	12,200	12,800	13,000	20,800	21,740	27,110	26,980	27,920	28,180	29,730	30,480	31,250
Yard Waste	20,000	23,200	27,500	35,000	29,690	30,530	27,980	31,160	31,470	31,770	32,070	32,400
Other**	1,370	2,550	4,770	6,090	6,800	7,690	7,820	7,860	7,940	8,110	8,230	8,270
Total MSW Generated*	**88,120**	**121,060**	**151,640**	**205,210**	**213,700**	**238,260**	**231,230**	**239,390**	**240,370**	**249,180**	**248,150**	**251,340**

*Generation before materials recovery or combustion. Does not include construction and demolition debris, industrial process wastes, or certain other wastes.

**Includes miscellaneous organic and inorganic wastes.

Source: U.S. EPA 2007

Table 5.2 Composition of Municipal Solid Waste Generated in the United States from 2000 to 2006 (in 1000 tons and percentage*)

Waste Component	Weight	Percentage	2006 Percentage
Paper and Paperboard	82,660–87,550	33.9–36.8	33.9
Glass	12,340–13,200	5.1–5.4	5.3
Metals	18,280–19,130	7.5–7.9	7.6
Plastic	25,270–29,490	10.6–11.7	11.7
Rubber and Leather	6,540–6,820	2.6–2.8	2.6
Textiles	9,810–11,840	4.0–4.7	4.7
Wood	13,020–13,930	5.5–5.7	5.5
Food Waste	26,980–31,250	11.4–12.4	12.4
Yard Waste	27,980–32,400	12.7–13.1	12.9
Other**	7,690–8,270	3.3–3.4	3.4
Total MSW Generated*	**231,230–251,340**	**100**	

*Generation before materials recovery or combustion. Does not include construction and demolition debris, industrial process wastes, or certain other wastes.
**Includes miscellaneous organic and inorganic wastes.
Source: U.S. EPA 2007

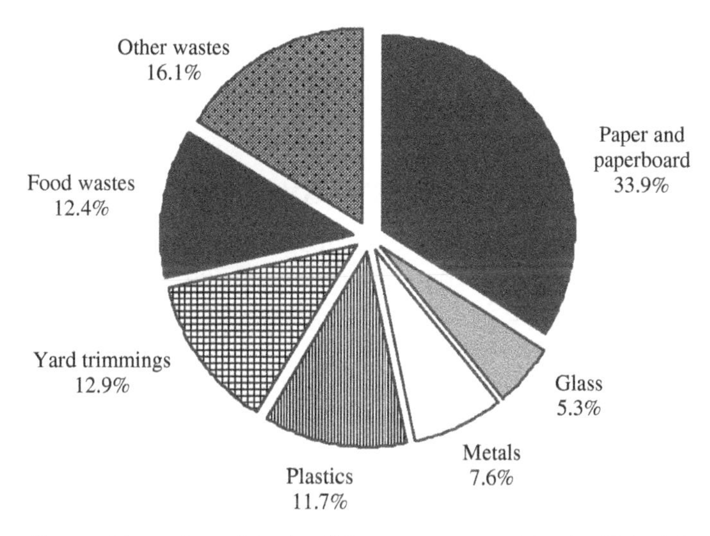

Figure 5.1 Composition of municipal solid waste generated in the United States in 2006 (percentages of total generation)

in 2006, although it also stabilized since 2004. Its low-density, high-strength, user-friendly design and fabrication capabilities and low cost are the drivers to such growth (Subramanian 2000). It is also interesting to note that the generation of glass, metals, and yard waste has been more or less steady throughout the years.

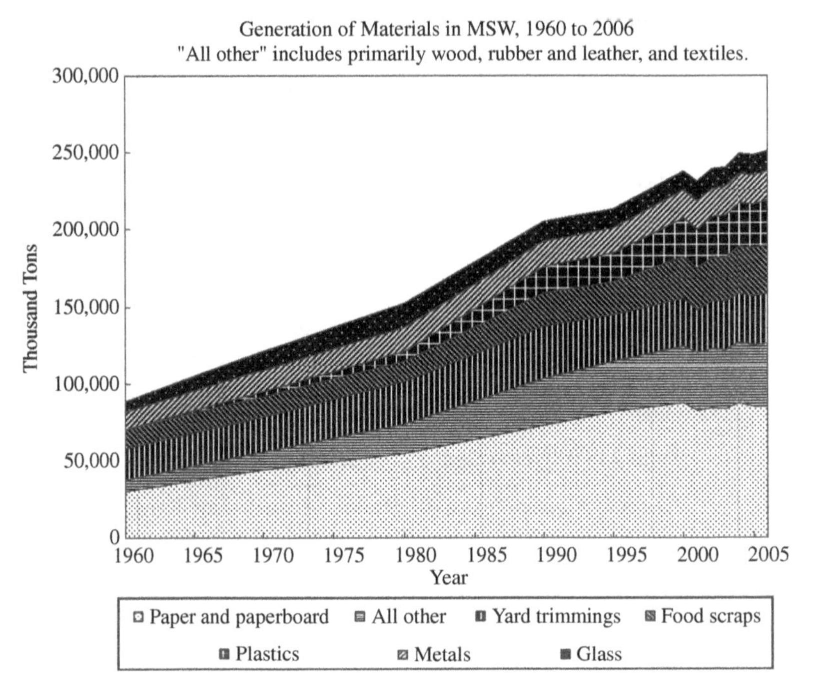

Figure 5.2 Municipal solid waste generated in the United States from 1960 to 2006 (U.S. EPA 2007)

3 SOLID WASTE MANAGEMENT SYSTEM

Based on the U.S. EPA tiered integrated waste management strategy, a proposed flow chart of a typical municipal solid waste management system is illustrated in Figure 5.3. The source of all municipal solid waste starts with consumers. During the past 45 years, the amount of waste each person generates has almost doubled, from 2.7 to 4.54 pounds per day. If the consumer and business can use less of a resource, it will be the best way to avoid the waste generation. However, this reduction of using resources involves education, and depends on business practices, various cultural, and social and economic factors that may not be controlled by the planner or engineer.

Source reduction, often called *waste prevention*, is defined by the U.S. EPA as "any change in the design, manufacturing, purchase, or use of materials or products (including packaging) to reduce their amount or toxicity before they become municipal solid waste. Prevention also refers to the reuse of products or materials." Source reduction can be an effective way to reduce the waste generated for collection and disposal. This includes reusing any jars and bottles, grocery bags, and waste papers. Consumer practices can also be changed to reduce waste. For example, consumers can purchase long-life products, repair used products instead of purchasing new ones, and reduce junk mail. Food and

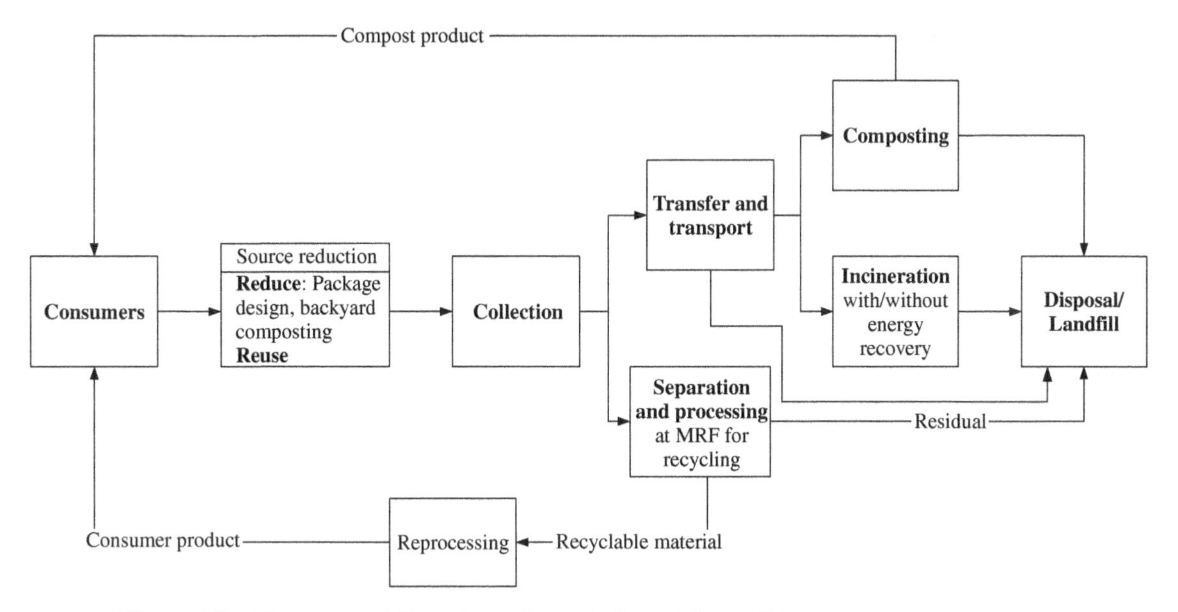

Figure 5.3 The proposed flow chart of a typical municipal solid waste management system

yard wastes can be reduced or eliminated by backyard composting. Industries can redesign products to reduce or eliminate waste. For example, they can lengthen the useful life of the product, use less materials in the product and packaging, use environmental friendly materials, design for secondary uses, and use paperless purchase orders.

If source reduction and reuse of products are not possible, recycling will be the next choice. Collecting recyclables varies from community to community, but there are four primary methods: curbside, drop-off centers, buy-back centers, and deposit/refund programs. Regardless of the method used to collect the recyclables, they are sent to a materials recovery facility (MRF) to be sorted and prepared into marketable commodities for manufacturing. More and more of today's products are being manufactured with total or partial recycled content. This is the last step in the recycling stage. Recyclables are bought and sold just like any other commodity, and prices for the materials will fluctuate with the market.

Solid waste collection and transportation can be described as follows: The waste is first put into waste bags or containers by individual households, after being collected by small waste-collection vehicles. The waste is transported to a transfer station for classification and pressing (transfer stations may not be needed if the disposal facility is close to the source). Waste can then be sent to a recycling center or an incinerator for resource and energy recovery. Finally, the residuals are taken by larger vehicles to a sanitary landfill or final disposal. The establishment of intermediate transfer stations where loose waste is pressed into a smaller volume and higher density and then loaded to larger truck to a disposal facility will help to reduce the costs for transportation. However, installation of these transfer stations will increase the cost accordingly. Separation of waste components is important in the collection of recyclables. While separation helps recycling of the useful resources, it also increases the cost for waste collection.

As discussed before, the collected recyclables or, in some cases, commingled solid waste are transported to the MRF for further separation and processing. The purpose of the MRF is to separate and recover materials and to improve the quality of recovered materials for sale in the market. The processes employed in an MRF are highly dependent on the types of waste or recyclables received and subsequent uses of the end products. They typically includes sorting, size reduction, screening, density separation, magnetic separation, compaction and baling.

As shown in Table 5.2, food waste and yard waste together constitute 25 percent of the MSW. Instead of going to landfills, this organic portion of the MSW can go through a composting process to produce compost for use as a soil conditioner for garden, farm, or landscape needs. Composting is a biological process where the organic material is decomposed with the help of microorganisms to a simpler organic mix that can be used as soil amendments. A critical step in the composting process is the destruction of pathogens and the U.S. EPA provides specific standards of pathogen control in composting systems.

Incineration has been used to reduce the residue of MSW in order to save the landfill space. Most of the incineration plants in the United States are mass-burned incinerator using MSW as collected for the fuel. Because the incineration process releases air pollutant and produces ashes and is expensive compared with other MSW disposals, it has been used only in areas where the landfill is not readily available. Almost all the recently built incineration plants have energy recovery facilities to offset the cost of operation and air pollution control. Because of the recent success of recycling effort, the energy content of MSW has been decreased; therefore, the demand of incineration with energy recovery has been significantly reduced.

Landfill has been a necessary disposal component in a MSW management system because there are significant residues produced in the recycling, composting, or incineration processes. A landfill is a highly regulated and specially engineered site for disposing of MSW on land. Because of the shortage of available land space and the fact that residents usually do not want a landfill in their communities, locating a new landfill becomes a difficult task. Landfill is an important component of the MSW management system because the waste diversion ratio is still much less than 50 percent. The trend of the MSW management is to place emphasis on the reuse and reduction of solid waste and to reduce the solid waste sent to a landfill before it goes through recycling, composting, or incineration process for resource recovery.

4 RECYCLING MUNICIPAL SOLID WASTE

Recycling is widely regarded to be environmentally beneficial and conducive to sustainable economic development. It saves our valuable resources, decreases demand for landfill space and generally saves energy in avoiding raw material extractions. Nevertheless, the collection, sorting and processing of recyclable materials for the manufacture of new products have their own environmental impacts. In addition, reduction in the amount of combustible substances such as paper and cloth, as a consequence of waste classification and recycling, results in a substantial decrease in the amount of heat generated during the course of waste incineration, and thus increases the incineration operating cost. Although incineration can lengthen the lifetime of landfills, the expenses are high for the construction, operation, and maintenance of incinerators. The typical solid waste for recycling is listed in this section.

4.1 Aluminum Cans

Because of the high price of aluminum and the aluminum industry's infrastructure support, the aluminum beer and soft drink can is the most recycled consumer beverage container in the United States (The Aluminum Association 2008). In 2006, 1.44 million tons of aluminum beverage cans were generated in MSW. Out of these, 0.65 million tons, or 45 percent, were recovered and 0.79 million tons were discarded. Used aluminum beverage can scrap is the major component of processed old scrap, which accounts for approximately one-half of the old aluminum scrap consumed in the United States. Due to improved production efficiency, the weight of an individual aluminum can has been decreased, and therefore, the number of aluminum cans produced by a pound of aluminum has increased. Aluminum beverage cans continue to make up the largest portion of the scrap aluminum purchased domestically. However, discarded aluminum products (old scrap) other than used beverage containers are also a significant source.

4.2 Paper and Paperboard

Paper and paperboard products constitute about 34 percent of the MSW stream. This is the largest portion of MSW. In 2006, Americans generated about 85 million tons of paper products, which are nearly a threefold increase from 1960. About 52 percent of all paper and paperboard products were recovered in 2006, nearly two and a half times the percentage in 1960 (U.S. EPA 2007).

4.3 Corrugated Boxes

Approximately 37 percent of paper and paperboard products in MSW are corrugated boxes. In 2006, 31.4 million tons of corrugated boxes were generated in MSW. Out of these 22.6 million tons, or 72 percent, were recovered and 8.8 million tons were discarded. The supply of corrugated boxes is from retail/commercial sources with 50 percent contribution, the manufacturing sector with 28 percent, residential at 13 percent, and pre-consumer supplies at 8 percent. Of the waste contributed by the retail/commercial sources, 75 percent is recovered (the manufacturing sector had 70 percent recovered, residential sources had 5 percent recovered), and finally, preconsumer supplies have nearly 90 percent recovered. The primary market for corrugated boxes is the paperboard industry, which uses corrugated boxes for corrugating medium, linerboard, recycled paperboard, and other paper products. There is increasing demand for corrugated boxes in the paperboard industry and this demand can only be met if the recovery rate can be increased from each of the supply sources.

4.4 Newspapers

Newspapers include newsprint and groundwood inserts. Newspapers recovered from the waste stream have a wide variety of applications. These applications include providing feedstock for a variety of recycled products such as newsprint, paperboard, tissue, containerboard, molded pulp, animal bedding, cellulose insulation, and a bulking agent for compost. In 2006, 12.4 million tons of newspapers were generated in MSW. Out of these, 10.9 million tons, or 87.9 percent, were recovered and 1.49 million tons were discarded. In general, roughly three quarters of the tonnage recovered was collected by local governments. The remaining portion came from the private sector.

4.5 Office-type Papers

Office-type papers are high-grade papers such as copy paper and printer paper, and are usually generated by offices. Other paper types can meet this definition but are generated and recovered from houses and other commercial facilities. Offices can also generate paper wastes that cannot be considered as office paper—for example, magazines and newspapers. In 2006, 6.32 million tons of office-type papers were generated in MSW. Out of these, 4.15 million tons, or 61.7 percent, were recovered and 2.17 million tons were discarded. The office paper generation during the 1990s remained almost flat due to the growth of electronic forms of information processing. The primary markets for recovered papers are tissue paper, new printing and writing (P&W) papers, and recycled paperboard. The main driver for sorted office paper demand is the strength of the de-inked pulp (DIP) market. The growing deinking facilities nationwide led to the increase of office paper consumption in recent years. Contamination also affects the successful production of DIP. Generally, more than one-third of sorted office paper exceeds the allowable levels of prohibited materials. To avoid such contamination, quality control is set at high levels.

4.6 Mixed Paper

Mixed paper includes discarded mail (third-class mail), telephone directories, catalogs, books, and magazines. Mixed paper may include all types of paper generated in offices and houses. Packages coated with plastics, such as frozen food and tissue containers, are not acceptable for recycling. Usually mixed papers may include other types of paper that are normally collected separately, such as office paper or old magazines. In 2006 10.3 million tons of mixed papers were generated in MSW. Out of these, 3.74 million tons, or 36.4 percent, were recovered and 6.53 million tons were discarded. In general, the private sector was responsible for more than 80 percent of the recovered quantity. Mixed paper is the fastest-growing recovered paper category in the past several years. This growth means that the industry is recovering a wide range of papers, but some paper grades may reach their maximum achievable levels quickly. Magazines and catalogs are collectively referred to as old magazines, since they are made of the same materials and are equally useful for end users. Old magazines, like other mixed paper, have traditionally been used as a low-grade paper supply for production of paperboard and tissue paper. Recently they have emerged as a valuable ingredient for recycled newsprint production, which has resulted in its collection separately from the recovered paper.

4.7 Plastics

The amount of plastics consumed annually in the United States has been growing steadily. Its generation in the MSW increased from less than 1 percent in 1960 to 11.7 percent in 2006, as can be seen in Table 5.1. In 2006, the United States generated 29.5 million tons of all plastics. Its improved strength, low density, user-friendly design and fabrication capabilities, and low cost are the drivers to such growth (Subramanian 2000). Plastics are a small but significant component of the waste stream. There are seven types of plastics coded from 1 to 7, but the major types of plastics now recycled in most communities are type 1, polyethylene terephthalate (PET), and type 2, high-density polyethylene (HDPE).

4.7.1 PET Plastic

One of the most popular resins used by the plastics industry is polyethylene terephthalate (PET), also known as polyester. It is extensively used in different variety of applications (e.g., plastic soda bottles). It is heavily used as polyester fiber in the manufacturing of clothing and carpeting. PET usage has grown rapidly due to the growth of soft drink container business recycled. The single-serve container is the fastest-growing market for PET bottles. According to U.S. EPA (2007), 3.06 million tons of PET were generated in MSW in 2006. Out of those, 0.62 million tons were recovered and 2.44 million tons were discarded. Soft drink bottles represent more than 30 percent of the PET generated (0.94 million tons) in the MSW. Recovery (0.29 million tons) would therefore target the largest portion of generated PET. For example, in North Carolina, recovered PET bottles are the primary source of PET recovery. Some municipalities in North Carolina stopped or slowed collection efforts due to low market prices. Most of the recovered PET material was recovered through local government programs. The contribution of the private sector in North Carolina for PET bottles recovery was very small (NCDENR 1998). Price and capacity are the main elements of PET market dynamics. They are very sensitive to fluctuations in virgin and off-spec markets. These fluctuations are directly related to international economic conditions and supply/demand balances. End users for recovered PET may include engineered resins, fiber, food and beverage containers, nonfood containers, sheet, film, and strapping.

4.7.2 HDPE Plastic

High-density polyethylene (HDPE) is obtained by polymerizing ethylene gas. The most common item that is manufactured from HDPE is milk jugs. Most of the current recovered HDPE is accomplished through local government collection programs. The most common form of recovered of HDPE is blow-molded bottles and HPDE grocery bags. The national generation of HDPE in MSW was 6.04 million tons in 2006. Out of these, 0.58 million tons were recovered and 5.46 million tons were discarded (U.S. EPA 2007). Milk and water bottles represent more than 10 percent of the PET generated (0.71 million tons) in the MSW. About 31 percent (0.22 million tons) of these bottles are recovered.

4.8 Steel Cans

According to U.S. EPA (2007), 2.75 million tons of steel packaging were generated in MSW in 2006. Out of those, 1.74 million tons, or 63.3 percent were recovered and 1.01 million tons were discarded. Food and other steel cans represent more than 90 percent of the steel package generated (2.51 million tons) in the MSW. The generation and recovery amounts are directly related to population growth.

It should be noted that steel cans represent a small portion, less than 20 percent, of the total ferrous metals (14.2 million tons) in MSW. Junked automobiles, demolished structures, worn-out railroad cars and tracks, appliances, and machinery are the major sources of obsolete scrap. The decrease in the percentage of the recycled steel is due to the increase in the production of durable steel products. The value of steel scrap is affected by the demand for finished products. As the demand expands, the need for more scrap steel will grow. The demand for steel is typically affected by the demand for cars. Efforts are underway to enhance the growth of other steel markets, and new technologies in steel production have increased the dependence on scrap. Nationally, the demand for steel can scrap is always more than the supply. The total demand for all steel scrap is much more than the part supplied through steel can recycling.

4.9 Glass

Glass is found in MSW primarily in the form of containers but also in durable goods such as furniture, appliances, and consumer electronics. In the container category, glass is found in beer and soft drink bottles, wine and liquor bottles, and bottles and jars for food, cosmetics, and other products. Most recovered glass containers (bottles) are used to make new glass containers, but a portion goes to other uses such as fiberglass insulation, aggregate, and glasphalt for highway construction. In 2006, 13.2 million tons of glass were generated in MSW. Out of these, 2.88 million tons, or 21.8 percent, were recovered and 10.3 million tons were discarded. Almost all the recovered glass tonnages were glass containers.

In general, most of the glass recovered was due to local government collection efforts. Although large quantities of glass were generated from commercial sources, quantities recovered from nonresidential locations were far below 10 percent of the total glass recovered. Most (more than half) of the glass waste can be characterized as flint (clear), followed by amber (brown). Less than 10 percent is green. Generally, the production of green glass is decreasing in the United States.

Glass containers marketers are classified into primary and secondary end users. Primary end users reuse the glass cullet (broken/crushed glass) to manufacture glass containers. Secondary end users use the glass for different purposes other than making glass containers. The glass container industry is the largest consumer for glass cullet in the United States.

Contamination is a major concern in glass recycling. The Institute of Scrap Recycling Industries specifications prohibit materials such as ferrous and nonferrous metals, ceramics, and other glass and other materials (bricks, rocks, etc.) from being present in glass cullet. Flint cullet must have no more than 5 percent nonflint cullet. Amber can withstand up to 10 percent nonamber cullet in the mix, and green can withstand up to 30 percent nongreen cullet. Many factors rather than economic factors affect supply and demand. The first and most important is public education. To meet high standards of glass is expensive and results in a low price paid for glass by the processor which makes it less profitable for generators. Transportation for long distances further increases the cost of glass recycling (NCDENR 1998).

5 MSW COLLECTION

Collection and transportation of solid waste is considered to be one of the most important stages of solid waste management because the cost of collection and transportation can reach as mush as 80 percent of all costs associated with solid waste removal. Thus, the optimization of waste collection and transportation services can yield large savings. As a result short term planning of vehicle routing and scheduling is valuable after the completion of long-term regional planning in a solid waste management system. Studies in this area concentrated in finding optimal truck routes and collection methods (Bhat 1996, Kulcar 1996, Movassaghi 1993, Teixeira et al 2004) as well as examining the insertion of certain processing facilities (such as transfer stations, power plants, etc.) between the waste source location and the final disposal destination, or in a general term the assignment of city zones to specific processing facilities or disposal sites for solid waste removal (Hsieh and Ho 1993, Chang and Wei 1999).

Optimization techniques can also be used in the planning of recycling drop-off stations and collection networks. Improving the cost-effectiveness of curbside collection schemes can be achieved by methods such as the distribution of recycling containers designed to store plastic, metal, glass, and paper. In 2006, more than 8,600 curbside recyclables collection programs were reported in the United States. The extent of residential curbside recycling programs varies tremendously by geographic region, with the most extensive curbside collection occurring in the Northeast (84 percent) and least in the South (30 percent). Overall, 46 percent of the population in the United States was served by curbside recyclables collection programs in 2006. Recycled materials are also collected by drop-off centers, buy-back centers, and through deposit systems. According to U.S. EPA (2005), ten states have container deposit systems: Connecticut, Delaware, Hawaii, Iowa, Maine, Massachusetts, Michigan, New York, Oregon, and Vermont. In these programs, the consumer pays a deposit on beverage containers at the point of purchase, which is redeemed on return of the empty containers. California has a similar system where containers can be redeemed, but the consumer pays no deposit.

6 MATERIALS RECOVERY FACILITIES (MRF)

There are generally two types of materials recovery facilities (MRF). One type that receives mixed solid waste (regular solid waste with recyclables mixed together) is called the *dirty MRF*. The other type that only receives separated recyclables is called the *clean MRF*. A typical dirty MRF consists of a large tipping floor with a material processing area and a storage area for different waste streams. All waste collected from the curbside is delivered to the tipping floor by the collection truck. The waste brought to the facility is therefore a combination of MSW garbage and recyclables. Easily separated recyclables such as cardboard are removed on the tipping floor before entering the processing center. Material is conveyed into a hopper from the tipping floor via bucket loader. There are a series of screens that separate out the smaller garbage from the recyclables. Recovery of recyclables can either be sorted automatically or manually. Manual sorting usually results in higher-quality materials with less downtime, but can be expensive because of the labor cost. Automated sorting is more effective for high throughput. Automated sorting equipment may include magnetic belts or drums for ferrous metal removal, eddy current separators for aluminum removal, and classifiers for separating light and heavy materials. Generally, a mix of manual and automated sorting is the most appropriate to ensure high-quality materials and minimize processing time. Recovery of recyclables at mixed-waste MRFs ranges from 15 to 20 percent of the input waste stream. A clean MRF is similar to a dirty MRF in that it has a tipping floor, processing, and storage areas. However, the recycled material is collected separately from the MSW, allowing for higher-quality material. There is less separation needed for the recyclables, compared to the dirty MRF. Most clean MRFs have two processing lines—one for commingled containers such as glass, plastic, and metal, and the second for fiber such as cardboard, newspaper, and high-grade paper. Similarly, clean MRFs can be manual or highly automated.

In 2005, 545 MRFs were operating in the United States, with an estimated total daily throughput of 86,000 tons per day (tpd). The distribution is 140 (24,351 tpd) in the Northeast, 156 (20,782 tpd) in the South, 132 (18,793 tpd) in the Midwest, and 117 (22,042 tpd) in the West.

7 COMPOSTING

Composting is a biological process where the material is decomposed with the help of microorganisms to a simpler organic matter that can be used as soil conditioners. The compost can be sold or given away to landowners, landscaping companies, and farmers for use in gardens, parks, and farms. Compost is environmental friendly and can reduce or eliminate the need for chemical fertilizers. It can also be used to remove solids, oil, grease, and heavy metals from stormwater runoff and amend contaminated, compacted soils (U.S. EPA 2007).

Here are the three most common composting methods:

1. *Aerated static pile composting*. This involves introduction of air into the stacked pile of mixed organic waste via perforated pipes and blowers. Layers of loosely piled wood chips or newspaper can also be added to facilitate the air passage from the bottom to the top of the pile. This method produces compost in three to six months. However, it is weather sensitive and thus may result in the loss of microorganisms responsible for the composting process.
2. *Aerated windrows composting*. Organics are formed into long and narrow piles that are turned with windrows turner equipment to reach required temperature and oxygen requirements. The pile height is between 4 and 8 feet which allows for the generation of sufficient heat and oxygen to flow to the center of the windrows. The pile width is usually between 14 and 16 feet. The disadvantage of this method is that it is used for only large volume of material thus requiring abundant space. There could be odor problems and leachate concerns with the application of this method.
3. *In-vessel systems*. Organic waste is fed into perforated barrels, silo drums, or specially manufactured containers that are simple to use and require

minimal labor. These equipments are not weather sensitive because the temperature, moisture and aeration are closely controlled. They vary in size and capacity although and they are usually used for handling small volumes of material. It only takes several weeks to produce the compost but takes several weeks to stabilize the compost. The initial cost of equipment setup may be high compared to other methods.

For small amount of household organic waste, two additional small-scale composting methods can be employed: backyard composting or vermicomposting. Backyard or onsite composting can take between one to two years to complete the composting process. However, occasional turning can improve the efficiency and reduce the composting period to six months. Vermicomposting uses red worms to mix with organic waste in composting bins to break down the organic wastes into a high-value compost called *castings*.

Several factors that influence the composting process include: carbon (C) to nitrogen (N) ratio, oxygen concentration, moisture content, particle size, temperature, and pH of the composting materials. Carbon and nitrogen are essential nutrients for the microorganisms to function properly in the composting process. The ideal C:N ratio should be 30:1. This ratio is important because a high C:N ratio could slow or halt the composting process. A low ratio, by contrast, could cause the organics to degrade too rapidly and use too much oxygen, causing unpleasant odors due to anaerobic conditions. It has been found that green, wet plant materials have a low C:N (high N), and brown, dry materials have a high C:N (high C). A proper blending of materials is thus necessary to achieve an appropriate ratio for the composting process.

Oxygen is needed for aerobic biodegradation, wherein microorganisms use oxygen to effectively degrade organic materials into carbon dioxide, humus, and inert mineral compounds. Without oxygen, the process becomes an anaerobic degradation process. In the anaerobic process, organic materials will still be degraded, but the process is relatively slow compared to the aerobic process, and it also causes unpleasant odors due to the presence of methane and noxious sulfur compounds. Water is an essential element for the composting process, helping to dissolve the organic and inorganic nutrients present in the compostable materials and making them available to soil microorganisms and their metabolic processes. The ideal moisture content of the compost pile should be between 40 and 60 percent by weight.

The surface area of organic materials exposed to soil organisms affects the rate of composting. The more finely ground a material, the higher the surface area per unit weight. Hence, large materials should be ground and shredded to smaller sizes. A mixture of materials should be used. When bulky materials are shredded in size, the decomposition rate increases while the porosity of the material decreases, resulting in anaerobic conditions. Ideal temperatures vary between 90° and 140°F. Maintaining high temperatures is necessary for rapid composting and destroying weed seeds, insect larvae, and potential plant and human pathogens. The temperature is measured with a long-stemmed thermometer at a depth of at least 18 inches into the volume of material collected. Temperatures above 140°F will begin to limit microbial activity, and temperatures in excess of 160°F can kill soil microorganisms. The pH of materials should be monitored, and a value 6.5 to 8 should be maintained. The role of bacteria in composting increases in importance with the increase in pH.

Factors to be considered in starting composting programs include available space and equipment, management personnel, capital and operation cost, and opportunity cost. A composting program could prove a viable alternative for management of yard waste and could help to reduce landfilled solid waste and disposal costs. Materials such as grass clippings and leaves collected through semiannual installation cleanups, waste from riding stables, and shredded classified documents could be useful for initiating a composting program. Dense materials such as trees and limbs are not suitable for composting due to their slow rate of decomposition, which will make the program expensive in operation as a result of long-term use.

Food waste is ideal for composting because of its high moisture content and susceptibility to odor production and large quantities of leachate. Fruits, vegetables, dairy products, grains, bread, unbleached paper napkins, coffee filters, eggshells, meats, and newspaper can be composted. Items unacceptable for composting include condiment packages, plastic wrap, plastic bags, foil, silverware, and drinking straws. Red meat, bones, and paper are acceptable, but they take longer time to decompose and thus are not preferred. Odor can be prevented by keeping the compost pile well aerated and free of standing water. Leachate can be reduced through aeration and by adding sufficient amounts of high-carbon bulking agent. Preconsumer food waste is easy to compose because it is generally separated from the rest of the waste stream generated, thus reducing the possible presence of contaminants in future compost. Postconsumer food waste is challenging because of separation issues involved, as the food waste is already mixed with general waste stream, thus increasing the presence of contaminants. This problem can be reduced by having a separate trashcan for only food waste.

8 INCINERATION

The major purpose of using incineration is to reduce the volume of the MSW so that the landfill space can be saved. Depending on the feed MSW composition, the volume can be reduced by 70 to 90 percent. Incineration is a thermal process used to convert the organic portion of the MSW to gases and the inorganic portion to ashes. There are two major types of incineration plants. One is mass-burn, which burns the commingled MSW, and the other is RDF-fired, which burns the refuse-derive fuel. There are three major issues with the incineration facilities: (1) the incineration plant is more expensive compared to other disposal facilities such as landfill; (2) it causes air pollution problem; and (3) the disposal of ash that may contain toxic metals. As with landfill, the siting of the facilities can be another problem.

Most of the municipal solid waste incineration currently practiced in this country incorporates recovery of an energy product (generally steam or electricity) at a *waste-to-energy facility* (WTE). The resulting energy reduces the amount needed from other sources, and the sale of the energy helps to offset the cost of operating the facility. Because of the success of the recycling program, it is important to predict and characterize the MSW feed to the WTE to provide enough waste and energy content of the waste for the generation of hot water or steam for heating and the generation of electricity.

The total amount of U.S. MSW incineration with energy recovery has decreased because of the success in recycling and reuse programs and the objections of communities surrounding the WTE facilities. In 2006, the design capacity of U.S. WTEs was 92,860 tons per day. There were 86 WTE facilities in 2006, down from 102 in 2000 and 112 in 1997. The distribution in 2006 is 40 (46,573 tpd) in the Northeast, 24 (31,131 tpd) in the South, 16 (10,912 tpd) in the Midwest, and 6 (4,280 tpd) in the West.

9 LANDFILLS

As mentioned before, landfills are a necessary disposal component in a MSW management system. A landfill is a specially engineered site for disposing of solid waste on land, constructed so that it will reduce hazard to public health and safety. The most important process in a landfill is the anaerobic decomposition of the organic portion of the MSW disposed in the landfill although aerobic decomposition occurs briefly in the beginning stage of landfilling when the air is available. The major products of the anaerobic decomposition are leachate and gases, including carbon dioxide (CO_2) and methane (CH_4). The leachate, a liquid containing high concentrations of organic compounds, can cause water contamination if not properly controlled. Both of these gases are greenhouse gases that cause global warming, although CH_4 is much more potent than CO_2. With a high energy content, CH_4 can be collected for energy production. If it is not harvested, it must be flared to reduce the environmental impact. Because the anaerobic decomposition process is relatively slow with limited moisture in the landfill designed under RCRA Subtitle D, leachate recirculation and landfill bioreactor have been developed to accelerate the organic decomposition rate in a landfill (Townsend et al. 1996, Mehta et al. 2002, Haydar and Khire 2006a 2006b, U.S. EPA 2006) and thus shorten the duration of the postclosure monitoring and management requirements.

The primary features of a landfill include an impermeable lower layer to block the movement of leachate into groundwater, a leachate collection system, a gravel layer permitting the control of methane, and a daily covering of waste with soil. These safeguards are intended to prevent groundwater contamination. However, it still possible that the system will fail and leachate will contaminate the groundwater by organics and heavy metals.

Solid waste disposal in landfills is highly regulated. The federal government, through the Resource Conservation and Recovery Act (RCRA), specified that landfills must meet certain minimum standards. In states whose RCRA Subtitle D programs have not been approved by the U.S. EPA, landfill must be designed with a composite liner, a leachate collection system, and point of compliance (POC) at the unit boundary. In states whose RCRA subtitle D programs have been approved by the U.S. EPA, the landfill design must be approved by state to ensure that maximum contaminant levels (MCLs) will not be exceeded in the uppermost aquifer at the POC, and the POC must be no more than 150 meters from unit boundary and must be on the property of the owner. There are also siting limitation for a landfill. For example, landfills must be 5,000 to 10,000 feet from an airport and it cannot be located in wetland, in a 100-year flood plain, or on fault lines. Landfill operations also must be extensively monitored by the operator through RCRA regulations. Additionally, the regulatory standards limit the amount of material that is allowed to be released into the groundwater and into the air.

Figure 5.4 shows the number of MSW landfills in the United States between 1988 and 2006. The number of MSW landfills decreased substantially over the past 18 years, from nearly 8,000 in 1988 to 1,754 in 2006. The distribution of these 1,754 landfills is 133 in the Northeast, 676 in the South, 425 in the Midwest and 520 in the West. Although the number of U.S. landfills has steadily declined

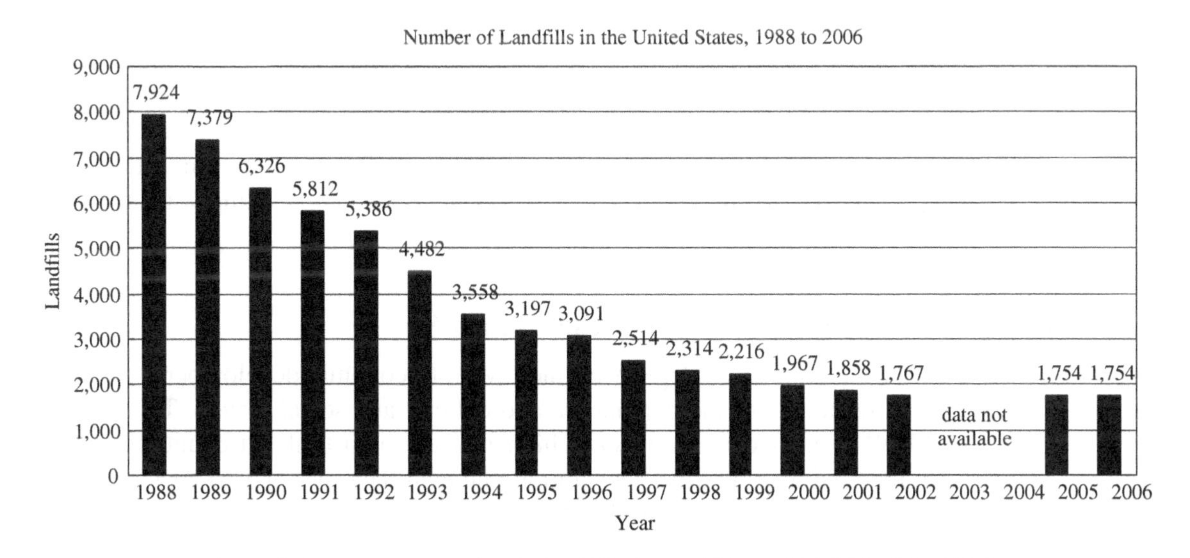

Figure 5.4 The number of U.S. landfills from 1988 to 2006 (U.S. EPA 2007)

over the years, the average landfill size has increased. At the national level, the total landfill capacity does not appear to be a problem, although regional shortages may sometimes occur.

Since 1990, the total volume of MSW going to landfills dropped by 4 million tons, from 142.3 million to 138.2 million tons in 2006. The net per capita discard rate (after recycling, composting, and combustion for energy recovery) was 2.53 pounds per person per day, down from 3.12 pounds per person per day in 1990 and the 2.63 pounds per person per day in 2000, and similar to the 2.55 per capita rate in 2004. The percentage of MSW that was landfilled increased slightly from 2005 to 2006. Over the long term, the tonnage of MSW landfilled in 1990 was 142.3 million tons, but decreased to 134.8 million tons in 2000. The tonnage declined to 133.3 million tons in 2005, then increased to 135.5 million tons in 2004.

10 SUMMARY OF HISTORICAL MSW MANAGEMENT

As already discussed, MSW generation has grown steadily from 88 million tons in 1960 to 251 million tons in 2006. Figure 5.5 shows the generation and recovery trend for the MSW management. Up to 1980, most of the MSW were landfilled, with a small percentage of recovery from recycling. After 1980, the significant effort in recycling and combustion with energy recovery reduced the amount of MSW disposal at landfills. In the meantime, recovery by composting also increased, although by a relatively smaller amount. If the population is taken into account, the per capita trend is shown in Figure 5.6. The per capita generation rate of MSW increased steadily from 1960 to 1990. The generation rate stabilized and remained at 4.5 lbs per capita per day from 1990 to 2006. The discard rate increased from 1960 to 1980 and started a decreasing trend after that because of the increased recycling rate. However, the recovery rate increase slowed and stabilized around 1.9 lbs per capita per day, and the discard rates stabilized around 2.6 lbs per capita per day. In 2006, 55 percent of the MSW is discarded, 45 percent was recovered. However, this recovery included 12.5 percent combustion with energy recovery. Thus, the recovery by recycling and composting was only 32.5 percent of the generated MSW.

11 MSW MANAGEMENT IN OTHER COUNTRIES

MSW generation rates depend on the level of economic development, environmental awareness of the population, and cultural and social factors. The amount of MSW generated is, in general, higher in the industrialized countries. It has been reported that the MSW generation rate is in the order of 1.6 kg (3.5 lbs) per person per day in the industrialized world and in the order of 0.8 kg (1.8 lbs) per person per day in Third World countries (Fehr et al. 2000). This rates are consistent with the data in the IPCC report (IPCC 2006) that list the MSW generation

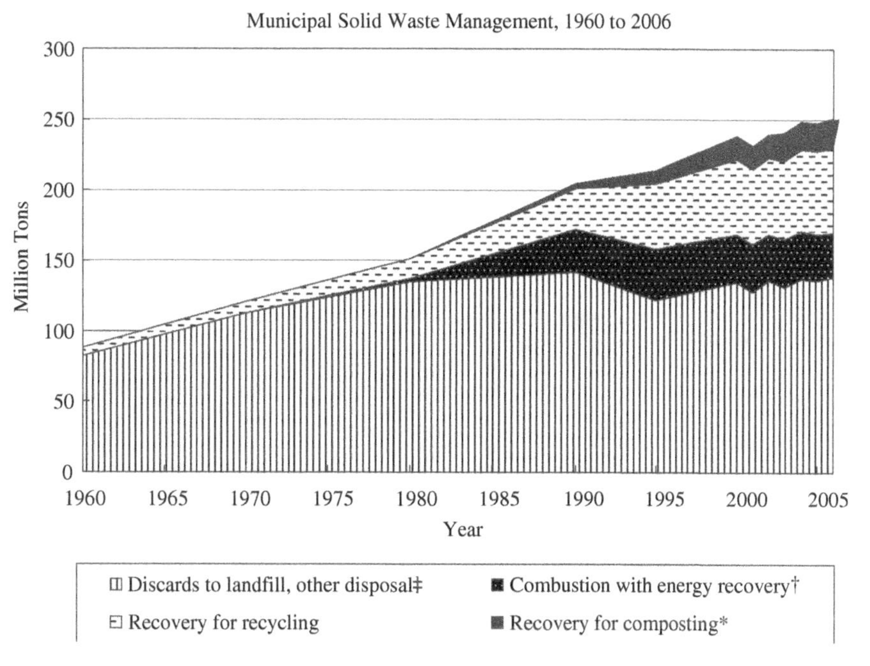

Figure 5.5 The MSW management trend in the United States (U.S. EPA 2007)

* Composting of yard trimmings, food scraps and other MSW organic material. Does not include backyard composting. (Neg. = Less than 5,000 tons or 0.05 percent)

† Includes combustion of MSW in mass burn or refuse-derived fuel form, and combustion with energy recovery of source separated materials in MSW (e.g. wood pallets and tire-derived fuel).

‡ Discards after recovery minus combustion with energy recovery. Discards include combustion without energy recovery. Details may not add to totals due to rounding.

rates as follows: Asia 1.2 to 2.0; Europe 2.0 to 3.5; Africa 1.5; North America 3.5; and South America 1.4 lbs/person/day. Furthermore, the characteristics of the MSW generated are also different. In the industrialized world, biodegradable material hardly reaches 50 percent of total generated household waste, while it usually between 60 to 80 percent in the Third World countries. Landfills are still viewed as the best waste disposal options in developing countries.

A comparative analysis of household waste in Stuttgart, Germany, and Kumasi, Ghana (Ketibuah et al. 2004), showed a distinctive contrast between a developed (Germany) and developing country (Ghana). Stuttgart is a typical city in a developed country where information and accurate data on MSW are documented, and as a result, it is easier to plan the collection and treatment of MSW. The MSW generation is about 1 kg per person per day, with a recycling rate of 38 percent. Incineration is the most used disposal method. Kumasi (Ghana) is a typical city in a developing country where there is no accurate documentation of MSW data, making it difficult to plan the collection and treatment of MSW for the years ahead. The daily waste generated per capita is estimated to be 0.6 kg.

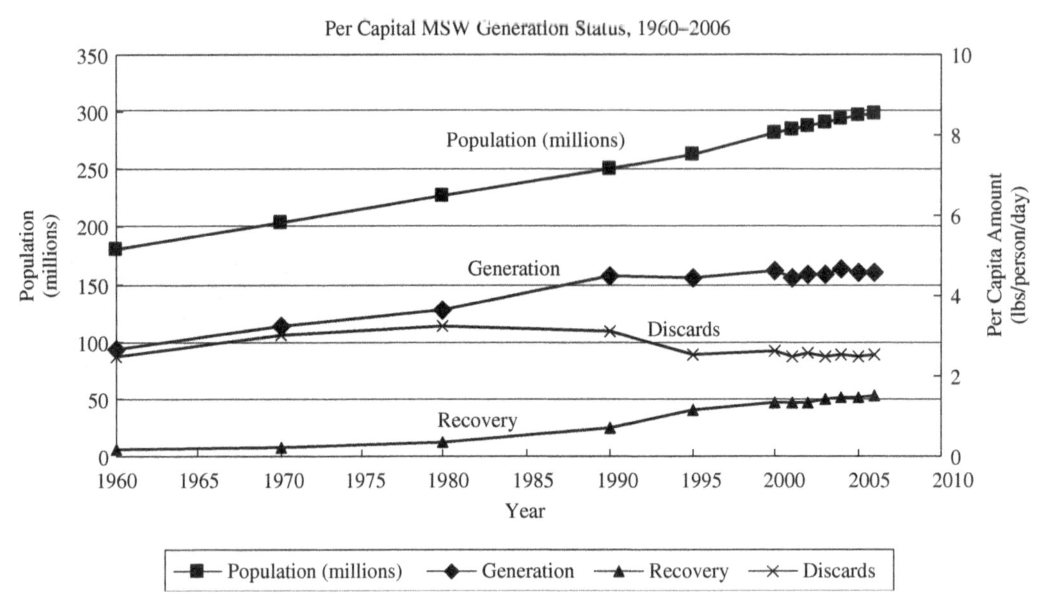

Figure 5.6 Per capita MSW generation, recovery and discards in the United States from 1960 to 2006 (U.S. EPA 2007)

Out of this amount, an estimated 70 percent is collected by private companies and dumped without pretreatment in a sanitary landfill. The rest (which is not collected) is dumped by individuals, usually in open spaces or in drains. The bulk of MSW was found to be organic waste, making composting a primary management option.

Table 5.3 shows the MSW recycling and disposal in selected European countries based on the selected yearly data from 1997 to 1999. It can be seen that the management of MSW varied significantly from country to country. For example, Greece, Portugal, United Kingdom, and Italy are heavily dependent on landfilling (91 to 78 percent), while Denmark, Netherlands, Austria, Belgium, and Sweden put less than one third (11 to 33 percent) of their MSW in the landfill. Denmark sent 50 percent of its MSW to incineration while Greece used none. Austria and Belgium have the best recovery (recycling and composting) rate, 64 and 52 percent, respectively, while Greece and Portugal have less than 10 percent of recovery. Even with 1997 to 1999 data, half of the European counties listed have a better MSW recovery rate (64 to 33 percent) than the United States (32.5 percent in 2006).

A more recent survey showed that in England, 67 percent of municipal solid waste (MSW) generated is landfilled, 9 percent incinerated, and 23.5 percent recycled or composted. During 2004/2005, 29.7 million tonnes of MSW was produced. This was 2.1 percent more than that produced in 2003/2004 (Husaini et al. 2007). The recycled or composted MSW increased to 27 percent during

Table 5.3 MSW Recycling and Disposal in Selected European Countries (in %)

Country	Landfilling	Incineration	Composting	Recycling	Unknown
Denmark (1999)	11	50	14	25	0
Netherlands (1999)	12	41	23	24	0
Austria (1999)	25	11	40	24	0
Belgium (1998)	27	21	15	37	0
Sweden (1998)	33	35	8	25	0
France (1998)	58	27	6	8	0
Spain (1999)	64	9	11	16	0
Finland (1997)	64	3	3	30	0
Italy (1997)	78	6	9	7	0
United Kingdom (1999)	81	8	2	9	0
Portugal (1999)	82	8	5	4	0
Greece (1997)	91	0	0	8	1

Source: Environmental Signals 2002, European Environment Agency (rounded to nearest whole percentage), Version 1, October 2003

2005/2006 (EFRA 2007). The average annual MSW increase in England was 1.5 percent from 2000/2001 to 2004/2005. Out of this, 25.7 million tonnes (about 86 percent) was from households alone. England has the capacity to engage in more recycling and composting of MSW (Curran et al. 2007). The EU Landfill Directive (1999/31/EC) requires a progressive reduction in biodegradable MSW to 75 percent of the 1995 disposal level by 2010 and 35 percent reduction by 2020. Various solid waste management schemes have been adopted in EU. These include the pay-per-bag scheme in Belgium and Italy, weight-based charging scheme in Denmark and Sweden, weight- and volume-based system in Germany and Luxemburg (Eunomia 2002, 2003); plastic bag environmental levy in Ireland (Dungan 2003), and other MSW management schemes (INFORM 2005, Green Alliance 2002). An empirical analysis of the effects of unit-based pricing of household waste for The Netherlands found that the weight- and bag-based pricing systems perform far better than the frequency- and volume-based pricing systems (Dijkgraaf and Gradus 2004).

European Council Directive (91/156/EC) urges the European member states to take appropriate measures to encourage the prevention and the reduction of waste production and its harmfulness (Gellynck and Verhelst 2007). It promotes the recovery of waste by means of recycling, reuse or reclamation, or any other process with a view to extracting secondary raw materials or the use of waste as an energy source.

What would induce a household to generate or throw away less waste (source reduction)? The answer hinges on at least two elements: the incentive built into the unit-pricing structure for waste collection and disposal, and the availability of convenient (and legal) alternatives such as recycling and yard waste collection or composting programs.

One of the major components of household waste is organic material such as kitchen and garden waste, typically comprising 43 percent by weight of an average household's waste in the Flemish region of Belgium. and may include vegetables, fruit, cooked and processed foods, weeds, grass, leaves, and other garden waste. The higher the annual average income of people in a municipality, the higher the amount of waste. A fortnightly collection of waste yields lower amounts of waste than a once-a-week collection round does (Gellynck and Verhelst 2007).

As mentioned before, developing countries in general produce less MSW. For example, the average amount of waste generated in Cuba is only 87 g per capita per day in the Santiago de Cuba Province (Mosler et al. 2006). The recycling rates are rather high, as the households have established different disposal strategies as a function of the waste type. Forty percent of the households feed organic waste to animals like pigs and chickens. Plastic, aluminum, and glass are separated at household level and stored until collection. The study of the municipal solid waste in Dakar (Senegal, Africa) indicated that solid waste management is a severe problem in big cities of developing countries (Kapepula et al. 2007). Their result tends to confirm a generally observed rule that people with higher income produce more garbage. Solid waste generation in the city of Mekelle, Northern Ethiopia has been increasing, however; per capita waste generation in Mekelle is estimated to vary between 0.30 kg/day and 0.33 kg/day (Tadesse et al. 2008). On average, only one-third has been collected and disposed. Solid waste collection service at household level by the municipality of Mekelle is primarily carried out using door-to-door collection services by tractor-trailers and collection service using fixed-point communal containers.

In Asia, the World Bank (1999) reported that residential solid waste represents about 30 percent of the overall municipal waste stream. It is estimated that the per capita waste generation rate in six major urban areas of Bangladesh is between 0.25 and 0.56 kg per day (Sujauddin et al. 2008). About 66 percent of this is compostable. The large organic concentration in urban solid waste indicates the necessity for frequent collection and removal. This also suggests the good potential for recycling of organic waste.

A number of socioeconomic variables may affect the quantity of solid waste generated each day by a household. These include religion, family size, family employment, age, education, land status, and duration of stay. Composition of solid waste depends on a number of factors, such as food habits, cultural traditions, socioeconomic status, and climatic condition. In 2005, the average daily amount generated per capita was 0.92 kg in Hangzhou, China (Zhuang et al. 2008). With a 4.1 million population, the total MSW is approximately 1.38 million tonnes and the average increase rate was about 3.3 percent over the past five years. About 60 percent of the MSW is disposed by landfill, and the remaining 40 percent by incineration. In comparison to the composition of MSW in other developed countries, the quantity of food waste in the studied communities is considerably higher (64 percent). The higher percentage and moisture content of food waste in household waste makes the separation of recyclable materials from the other waste difficult. The quantity of potentially recyclable materials is relatively lower. It correlates mainly with the dietary habit of the Chinese people, who prefer food that is unprocessed and unpackaged.

12 MSW PLANNING ISSUES AND OPTIMIZATION MODELS DEVELOPMENT

The municipal solid waste disposal is a growing problem throughout the world as a result of economic development and population growth. This problem is especially serious in densely populated metropolitan areas. Thus, it is critical for the planners and engineers in locating appropriate number and type of solid MSW processing and disposal facilities or their expansion for the proper handling of MSW. Because solid waste disposal facilities are potential sources of environmental and health risks (Hagemeister et al. 1996), these facilities are considered undesirable by many communities and organizations because residents do not want such facilities to be located near them. Thus, to locate a solid waste disposal facility, not only the technical and economical requirements have to be satisfied, but social and political issues also have to be considered.

In general, landfill construction occurs in phases, where sections of the available area at developed sequentially. It's also useful to know that landfills are designed with decades of capacity, while the modules within a landfill typically have operating lives from one to several years. Most of the studies in the literature are related to landfills and disposal facilities focused on facility siting, and this could be attributed to the numerous landfills in the United States that were reaching their capacity or were about to be closed due to environmental regulations. It was also becoming harder and harder to site new landfills. Some states at the time had as few as 3 years of remaining landfill capacity, while others had as many as 20 years.

Models for integrated municipal solid waste capacity planning can provide valuable insight into trade-offs between landfill diversion program costs and savings from resulting landfill service lifetime extensions (Lund 1990, Lawver and Lund 1995). Some of the studies in this area focused on finding the optimal size when building a new landfill, while other studies discussed the expansion of an existing landfill (Huang et al.1995).

Another concern with landfill siting is the conflicting objectives and constraints, because even though a potential site for a landfill may have a good geological formation that would impede ground water contamination from the landfill, it could be deemed unacceptable because it is near a housing development or other areas planned for public or private use (Minor and Jacobs 1994). MSW management facilities siting was also included in studies that focused on solid waste management on a regional level where the interaction between various parameters and factors such as system costs and scheduling has been illustrated (Movassaghi 1993; Hsieh and Ho 1993; Lund 1990; Lund et al. 1994). Uncertainty with respect to model parameters, as well as objectives, has also been explored with various optimization approaches (Chang et al. 1993; Huang, et al. 1995; Huang et al. 1997; Chang and Wang 1997).

The need for recycling has significantly increased with the rapid depletion of landfills throughout the 1980s along with the continuous decrease in many natural resources, and that lead to many federal and state regulations that recognized and enforced recycling in many industries. The fact that recycling is a costly solid-waste operation created a need for effective recycling programs with cost and environmental balances. The majority of the studies associated with recycling and composting were done in comparison with landfill and incineration. For example, examining the effect of increasing the efficiency of waste disposal through implementing recycling programs in regional levels by considering the cost of transportation and profits from the recycled materials (Everett et al. 1993; Keeler and Renkow 1994; Vogtlander et al. 2001).

There are also comprehensive models developed for general system-level simulation and cost estimating for components of solid waste management systems. Examples include the following:

- Anex et al. (1996) developed GIGO, a spreadsheet-based model of municipal solid waste management systems.
- Berger et al. (1999) developed EUGENE, a process-oriented optimization model that deals with the integrated solid waste management planning problem at a regional level.
- Wang et al. (1994) developed SWIM, an optimization model that provides decision support for the design of integrated solid waste management systems with the consideration of cost effectiveness and environmental impact.
- MacDonald (1996) developed a specific spatial decision support system (SDSS) by providing analytical tools for developing plans and evaluating a number of impacts associated with a plan to address the multiattribute and geographical nature of solid waste systems.
- Chang and Li (1997) presented a model utilizing modeling-to-generate-alternatives (MGA) for generating preliminary solid waste management (SWM) alternatives.
- Dowie et al. (1998) developed and implemented an institutional solid waste environmental management system (SW-EMS) in Pinawa-Canada. Several audits before and after the use of the system suggest that SW-EMS was successful in significantly reducing the waste sent to landfills.
- McGrath (2001) introduced a software tool called SMARTWaste that has been used to audit, reduce, and target waste arising on a construction site.

- Solano et al. (2002a, 2002b) developed a comprehensive life-cycle-based model for solid waste management. The model can be used to identify solid waste management alternatives with the consideration of cost, energy consumption, and environmental emissions.

13 ENVIRONMENTAL IMPACT OF MSW AND LIFE-CYCLE ASSESSMENT

The public and private sectors in the United States have come to realize the importance of addressing the long-term consequences of solid waste management. Providing our society with goods and services generates MSW and the associated environmental impacts. For example, anaerobic processes in a landfill generate leachate and gases, including carbon dioxide (CO_2) and methane (CH_4). The leachate can cause water contamination, and both gases cause global warming. Although the U.S. EPA estimated that roughly 60–90 percent of the methane emitted from the landfill can be captured by a landfill gas extraction project, only less than 20 percent of the landfills in the United States have actually implemented the extraction project (U.S. EPA 2008). Incineration, by contrast, causes air pollution and the disposal problems of the fly ash and bottom ash that may contain toxic metals. Sustainable development requires methods and tools to measure and compare these environmental impacts. Life-cycle assessment (LCA) has been used in the past as a tool for examining the environmental impacts of specific products. The purpose of life-cycle assessments (LCA) is to provide an understanding of the environmental impacts associated with a product, process, or activity. Life-cycle assessment is made up in four stages, which are used to calculate the environmental impacts for newsprint, office paper, and paperboard. The four major stages of the LCA are the goal definition, inventory, impact assessment, and valuation assessment (Craighill and Powell 1996).

The use of LCA in solid waste management has resulted in evaluating and comparing alternative disposal systems in regards to minimizing the environmental impacts associated with disposal techniques. Arena et al. (2003) adopted a life-cycle inventory aimed at identifying and quantifying the environmental burdens or impacts. The results indicate that recycling of solid waste is always environmentally preferable and that energy recovery from the processed materials is possible. Mendes et al. (2003) employed LCA and assessed the environmental impact of landfilling, composting, and biological treatment for the biodegradable fraction of MSW.

LCA was employed for the evaluation of technology such as recycling as an alternative to existing landfilling practices. Craighill and Powell (1996) indicated that landfilling contributed to greater global warming of greenhouse gas emissions than recycling. Nakamura (1999) observed that recycling is an effective way of achieving a reduction in the amount of waste and emissions. Recycling reduces the CO_2 emissions even when the efficiency of collection of waste is low, which was not the case for the other processes such as landfilling or incineration. It has also been reported that, in general, recycling is a better option than landfilling with respect to environmental impacts (global warming gases, acidification potential and nitrification potential) and energy consumption (Chang and Bindiganavile 2005a 2005b, Chang Elobeid 2005, Chang and Chisolm 2007).

Molgaard (1995) describes ecoprofiles—assessment of environmental and resource impacts—for six different ways of disposing the plastic fraction in MSW: two different material recycling processes involving the separation of plastic waste; material recycling without separation of plastic waste; pyrolysis; incineration with heat recovery; and landfilling. The results indicate that recycling of plastic from MSW is only environmentally and resource sound if it is separated into its generic plastic types, which makes it possible to produce a recycled plastic with properties comparable to virgin plastic. Alternatively, if it is not possible to separate the plastic, incineration with heat recovery is the most environmentally and resource-sound process. Beigl and Salhofer (2004) compare the ecological effects and costs of different waste management systems for individual waste types—waste paper, plastic packaging, metal packaging, and waste glass—by means of a life-cycle assessment and a cost comparison.

The results indicate that the waste paper recycling scenario has clear energy savings and reduces acid emissions with a similar cost and amount of global warming emissions. The recycling of waste glass and metal packaging leads to clear ecological benefits, whereas recycling of plastic packaging is ecologically advantageous, although it causes much higher collection and treatment costs.

Multicriteria methods have been used in evaluating and comparing the various technologies for treatment and disposal of solid waste. They are particularly useful in the valuation stage in the life-cycle assessment. The commonly used methods in evaluating the various disposal systems include the following:

- *Weighted aggregate method*. This method consists of identifying a list of alternative courses of actions and a set of evaluative criteria, as well as weights that reflect the importance of the criteria. Further, a dominance pairwise comparison of the alternatives wherein each alternative is evaluated against the other is performed, or the scores are finalized by comparing the arithmetic total of the scores of each of the alternatives in determining the best alternative solution. This method has been used in the selection of solid waste management and disposal options such as landfilling, waste to energy, recycling, composting, and incineration (Chung and Poon 1996; Powell 1996).
- *Goal programming method*. The goal programming method is designed to combine the logic of optimization in mathematical programming with the decision maker's desire to satisfy several goals. Two main types of goal programming are recognized: preemptive (lexicographic) and nonpreemptive (weighted) goal programming. Solution techniques of both types of

goal programming focus on the minimizations of the deviations from the target values of each goal subject to both goal constraints and original functional constraints. Chang and Wang (1997) propose a goal-programming model to evaluate the compatibility issues between MSW recycling and incineration. It addresses the goals of economic efficiency (cost/benefit analysis) and environmental protection involved in the solid waste collection, recycling, and treatment tasks (combustions temperatures and recycling ratios).

- *Outranking methods*. Electre and Promethee are the commonly used outranking decision-aid methods in solving various environmental problems. These outranking methods have been recommended for situations where there is a finite number of discrete alternatives. An important advantage of outranking methods is the ability to deal with ordinal and more or less descriptive information on the alternative methods or processes to be evaluated. Furthermore, the uncertainty concerning the values of the criterion variables can be taken into account using fuzzy relations, determined by indifference and preference thresholds. The main drawback of these methods is the interpretation of the results obtained. In the Electre methods, there is no need to turn the outcomes into monetary dimensions. The method provides the possibility to take into account the preferences of decision makers and any other external factor to be considered in the evaluation. Electre II method (Hokkanen et. al. 1995) and Electre III method (Karagiannidis and Moussiopoulos 1997; Hokkanen and Salminen 1997; Hokkanen, et al. 1995) were used to identify the best municipal solid waste management system among incineration, RDF combustion, and landfilling for a region.

14 FUTURE TRENDS IN MSW MANAGEMENT

It is ironic that while the MSW is buried in the landfill or burned in the incinerator, the manufacturers extract new raw materials to make some of the same products (e.g., plastics and paper products). All MSW management alternatives are associated with costs, and the cost is expected to go up. Although in the past the MSW management strived to select the best alternatives to managing the generated waste—with the awareness of environmental impact of the MSW management and the pursuit of sustainable development—reduction by waste prevention and reuse should be the first priority for MSW management. Unfortunately, while the U.S. EPA did encourage reduce, reuse, and recycling as the tiered integrated waste management strategy, RCRA and other federal guidelines are concentrated on the safe design of disposal facilities, especially the landfill design. Therefore, the overall MSW generation rate has not started a decreasing trend yet in the United States. European countries are usually more aggressive in waste reduction and recycling. For example, a recent presentation in the United

Kingdom proposed a comprehensive strategy to reduce waste (EFRA 2007). Furthermore, zero waste concept has been proposed and promoted by various groups (Connett 2000; Zero Waste Alliance 2008). *Zero waste* promotes waste prevention, reuse, and recycling. In this context, it is not much different from the U.S. EPA's tiered integrated waste management strategy. However, the goal of zero waste is that all waste has to be recycled into the marketplace or nature so that human health and environment are protected.

The management of MSW is a very complicated task for the government as well as the public. Short-term solutions must be balanced with the long-term consequences of solid waste management. For example, with the emphasis of waste reduction, reuse, and recycle, the use of landfill and incineration in the MSW management needs further and extensive evaluation. Locating new MSW facilities will be a challenging task for the planners and engineers. Conflicting interests such as environmental impacts, economical efficiency, and social justice must be considered in choosing the management alternatives and locating MSW facilities. Life-cycle analysis has been a useful tool to characterize the total environmental and energy impact, while the inclusion of local communities and environmental groups in the planning process can make the MSW management implementation more feasible.

Acknowledgment

I would like to express my appreciation to Wanchi Huang for her assistance in most of the data collection and in providing tables and figures based on those data. I also benefited from the research conducted by my graduate students in solid waste management for the past several years.

REFERENCES

The Aluminum Association. http://www.aluminum.org. Accessed May 2008.

Anex, R., R. Lawver, J. Lund, and G. Tchobanoglous. 1996. GIGO: Spreadsheet-based simulation for MSW systems. *Journal of Environmental Engineering* 122 (4).

Arena, U., M. L. Mastellone, F. Perugini. 2003. Life cycle assessment of a plastic packaging recycling system. *International Journal of Life cycle analysis* 8(2): 92–98

Berger, C., G. Savard, and A. Wizere. 1999. EUGENE: An optimization model for integrated regional solid waste management planning. *International Journal of Environment and Pollution* 12 (2/3).

Beigl, Peter, and Stefan Salhofer. 2004. Comparison of ecological effects and costs of communal waste management systems. *Resources Conservation and Recycling* 41: 83–102.

Bhat, V. 1996. A model for the optimal allocation of trucks for solid waste management, *Waste Management & Research* 14: 87–96.

Chang, Ni-Bin, and S.F. Wang. 1997. Integrated analysis of recycling and incineration programs by goal programming techniques. *Waste Management and Research* 15: 121–136.

Chang, N., and Y. Wei. 1999. Strategic planning of recycling drop-off stations and collection network by multiobjective programming. *Journal of Environmental Management* 24 (2): 247–263.

Chang, N., R. Schuler, and C. Shoemaker. 1993. Environmental and economic optimization of an integrated solid waste management system. *Journal of Solid Waste Technology and Management* 21 (2).

Chang, S. Y., and Z. Li. 1997. Use of a computer model to generate solid waste disposal alternatives. *Journal of Solid Waste Technology and Management* 24 (1).

Chang, S. Y., and K. Bindiganavile. 2005. Environmental impacts of municipal solid waste recycling: A case study. *Proceedings of the 20th International Conference on Solid Waste Technology and Management*, Philadelphia, PA, April 3–6 2005.

Chang, S. Y., and E. A. Elobeid. 2005. Life cycle assessment of plastics recycling. *Proceedings of the 20th International Conference on Solid Waste Technology and Management*, Philadelphia, PA, April 3–6 2005.

Chang, S. Y., and K. Bindiganaville. 2005. LCA and multicriteria evaluation of waste recycling. *Proceedings of the 4th International Conference on Environmental Informatics*, Xiamen, China, July 26–28, 2005.

Chang, S. Y., and K. Chisolm. 2007. The environmental effects of recycling and manufacturing paper. *Proceedings of the 22nd International Conference on Solid Waste Technology and Management*, Philadelphia, PA, March 19–21.

Chang, S. Y., R. Cramer, K. Bindiganavile, and E. Elobeid. 2008. Locating regional materials recovery facilities: A case study. *Proceedings of the 23nd International Conference on Solid Waste Technology and Management*, Philadelphia, PA, March 30–April 2.

Chung S. S., and C. S. Poon. 1996. Evaluating waste management alternatives by multiple criteria approach, *Resources, Conservation and Recycling* 17: 189–210.

Connett, P. 1998, Municipal waste incineration: A poor solution for the twenty first century, Presentation at the 4th Annual International Management Conference Waste-To-Energy, Nov. 24 to 25, Amsterdam. http://www.mindfully.org/Air/Connett-Waste-Incineration-24nov98.htm

Connett, P. 2000. Talk given at the launching of Target Zero Canada, Toronto, Canada, November 21, http://www.grrn.org/zerowaste/articles/paul_connett_tzc_11-21-00.html.

Craighill, A. L., and J. C. Powell. 1996. Life-cycle assessment and economic evaluation of recycling: A case study, *Resources, Conservation and Recycling* 17: 75–96.

Curran, A., Williams, I. D., and S. Heaven. 2007. Management of household bulky waste in England. *Resources, Conservation and Recycling*, 51: 78–92.

Daniel, H.; L. Thomas, 1999. *What a waste: Solid waste management in Asia*. The World Bank, Urban & Local Government Working Paper Series No. 1, Washington, DC.

Dijkgraaf, E., and R.H.J.M. Gradus. 2004. Cost savings in unit-based pricing of household waste. The case of The Netherlands. *Resource and Energy Economics* 26: 353–371.

Dowie, W., D. McCartney, and J. Tamm. 1998. A case study of an institutional solid waste environmental management system, *Journal of Environmental Management* 53: 137–146.

Dungan L. 2003. What were the effects of the plastic bag environmental levy on the litter problem in Ireland? http://www.colby.edu/personal/t/thieten/litter.htm.

Environment, Food and Rural Affairs. 2007. Waste strategy for England 2007, http://www.defra.gov.uk/. Accessed June 2008.

Eunomia. 2002. Financing and incentive schemes for municipal waste management case studies. Final Report to Directorate General Environment, European Commission.

Eunomia. 2003. Eurocharge: Charging schemes for waste management and the barriers to their introduction in the UK. Final report to IWM (EB), Northampton, UK.

Everett, J., A. Modak, and T. Jacobs. 1993. Optimal scheduling of composting, recycling, and landfill operations in an integrated solid waste management system. *Journal of Solid Waste Technology and Management* 21 (3).

Fehr, M., M.S.M.V. de Castro, and M.d.R. Calcado. 2000. A practical solution to the problem of household waste management in Brazil. *Resources, Conservation and Recycling* 30: 245–257.

Gellynck, X., and P. Verhelst. 2007. Assessing instruments for mixed household solid waste collection services in the Flemish region of Belgium. *Resources, Conservation and Recycling* 49: 372–387.

Green Alliance. 2002. Creative policy packages for waste: Lessons for the UK http://www.greenalliance.org.uk/.

Hagemeister, M., D. Jones, and W. Woldt. 1996. Hazard ranking of landfills using fuzzy composite programming. *Journal of Environmental Engineering* 122 (4).

Haydar, M.M., and M. Khire. 2006a. Leachate recirculation using permeable blankets in engineered landfills. *Journal of Geotechnical & Geoenvironmental Engineering, ASCE* 133(2): 166–174.

Haydar, M., and M. Khire. 2006b. Leachate recirculation using horizontal trenches in bioreactor landfills. *Journal of Geotechnical & Geoenvironmental Engineering, ASCE* 131(7): 837–847.

Hokkanen, Joonas, and Pekka Salminen. 1997. Choosing a solid waste management system using multicriteria decision analysis, *European Journal of Operational Research* 98: 19–36.

Hokkanen, J., P. Salminen, E. Rossi, and M. Ettala. 1995. The choice of a solid waste management system using the ELECTRE II decision-aid method. *Waste Management and Research* 13: 175–193.

Hsieh, H., and K. Ho. 1993. Optimization of solid waste disposal system by linear programming technique. *Journal of Solid Waste Technology and Management* 21: (4).

Huang, G., B. Baetz, and G. Patry. 1995. Grey integer programming: An application to waste management planning under uncertainty, *European Journal of Operational Research* 83: 594–620.

Huang, G., B. Baetz, G. Patry, and V. Terluk. 1997. Capacity planning for an integrated waste management system under uncertainty: A North American case study. *Waste Management & Research* 15.

Husaini, I. G., A. Garg, A., K. H. Kimb, J. Marchant, S.J.T. Pollard, and R. Smith. 2007. European household waste management schemes: Their effectiveness and applicability in England. *Resources, Conservation and Recycling* 51: 248–263.

INFORM. Executive summaries. INFORM: strategies for a better environment 2005. http://www.informinc.org/.

IPCC. 2006. Guidelines for National Greenhouse Gas Inventories, Chapter 2: Waste Generation, Composition and Management Data, http://www.ipcc-nggip.iges.or.jp/public/2006gl/pdf/5 _Volume5/V5_2_Ch2_Waste_Data.pdf, accessed June 2008.

Kapepula, K.-M., G. Colson, K. Sabri, and P. Thonart. 2007. A multiple criteria analysis for household solid waste management in the urban community of Dakar. *Waste Management* 27: 1690–1705.

Karagiannidis, Avraam, and Nicolas Moussiopoulos. 1997. Application of Electre III for the integrated management of municipal solid wastes in the Greater Athens area, *European Journal of Operational Research* 97, 439–449.

Keeler, A., and M. Renkow. 1994. Haul trash or Haul ash: Energy recovery as a component of local solid waste management. *Journal of Environmental Economics and Management* 27.

Ketibuah, E., M. Asase, S. Yusif, M. Y. Mensah,. and K. Fischer. 2004. Comparative analysis of household waste in the cities of Stuttgart and Kumasi—Options for waste recycling and treatment in Kumasi. Proceedings of the 19th International CODATA Conference, Berlin, Germany. http://www.codata.org/ 04conf/papers/Ketibuah-paper.pdf.

Kim, J., and F. Pohland. 2003.. Process enhancement in anaerobic landfill bioreactors. *Water Science and Technology* 48(4): 29–36.

Kulcar, T. 1996. Optimizing solid waste collection in Brussels. *European Journal of Operational Research* 90.

Lawver, R., and J. Lund. 1995. Least cost replacement planning for modular construction of landfills, *Journal of Environmental Engineering* 12(3).

Lund, J. 1990. Least cost scheduling of solid waste recycling, *ASCE Journal of Environmental Engineering* 116 (1).

Lund, J., R. Anex, R. Lawver, and G. Tchobanoglous. 1994. Linear programming for analysis of material recovery facilities, *Journal of Environmental Engineering* 120 (5).

Mata, Teresa M., and Carlos A. V. Costa. 2001. Life cycle assessment of different reuse percentages for glass beer bottles, *International Journal of Life Cycle Assessment*, 6 (5): 307–319.

MacDonald, M. 1996. Solid waste management models: A state of the art review. *Journal of Solid Waste Technology and Management* 23 (2).

McGrath, C. 2001. Waste minimization in practice. *Resources, Conservation and Recycling* 32: 227–238.

Mehta, R., M. A. Barlaz, R. Yazdani, D. Augenstein, M. Bryars, and L. Sinderson. 2002. Refuse decomposition in the presence and absence of leachate recirculation. *J. Environmental Engineering* 128(3): 228–236.

Mendes, M. R., T. Aramaki, and K. Hanaki. 2003. Assessment of the environmental impact of management measures for the biodegradable fraction of municipal solid waste in Sao Paulo city. *Waste Management* 23: 403–409.

Minor, S., and T. Jacobs. 1994. Optimal land allocation for solid waste and hazardous waste sitting. *Journal of Environmental Engineering* 120 (5).

Molgaard, Claus. 1995. Environmental impacts by disposal of plastic from municipal solid waste. *Resources Conservation and Recycling* 15: 51–63.

Mosler, H. J., S. Drescher, C. Zurbrugg, C., T.C. Rodriguez, and O.G. Miranda, 2006. Formulating waste management strategies based on waste management practices of households in Santiago de Cuba, Cuba. *Habitat International* 30: 849–862.

Movassaghi, K. 1993. Optimality in a regional waste management system. *Journal of Solid Waste Technology and Management* 21 (3).

Nakamura, S. 1999. An interindustry approach to analyzing economic and environmental effects of the recycling of waste. *Ecological Economics* 28: 133–145.

North Carolina Department of Environment and Natural Resources. 1998. 1998 North Carolina Markets Assessment of the Recycling Industry and Recyclable Materials.

North Carolina Division of Waste Management 2003., Solid Waste Permits, http://www.p2pays.org/ref/01/00494.htm.

Powell, C. Jane. 1996. The evaluation of waste management options. *Waste Management and Research* 14.

Sciubba, Enrico. 2003. Extended exergy accounting applied to energy recovery from waste: The concept of total recycling, *Energy* 28: 1315–1334.

Solano, E., R. S. Ranjithan, M.A. Barlaz, O. P. Kaplan, and D. E. Brill. 2002a. Life-Cycle-based solid waste management. I: Model Development, *Journal of Environmental Engineering* 128 (10).

Solano, E., R. D. Dumas, K.W. Harrison, R.S. Ranjithan, M.A. Barlaz, and D. E. Brill. 2002b. Life-Cycle-based solid waste management. II: Illustrative applications, *Journal of Environmental Engineering* 128 (10).

Subramanian, P. 2000. Plastics recycling and waste management in the US, *Resources, Conservation and Recycling* 28 (3–4): 253–263.

Sujauddin, M., S.M.S. Huda, and A.T.M. R. Hoque. 2008. Household solid waste characteristics and management in Chittagong, Bangladesh. *Waste Management*. 28 (9).

Tadesse, T., A. Ruijs, and F. Hagos. 2008. Household waste disposal in Mekelle City, Northern Ethiopia. *Waste Management*. 28 (10).

Tchobanoglous, G., H. Thiesen, and S. Vigil. 1993. *Integrated solid waste management: engineering principles and management issues*. Hightstown, NJ: McGraw-Hill.

Teixeira, J. T., A. A. Pais, and D. S. Jorge Pinho. 2004. Recyclable waste collection planning—A case study. *European Journal of Operational Research* 158 (3).

Townsend, T., W. Miller, H. Lee, and J. Earle. 1996. Acceleration of landfill stabilization using leachate recycle. *Journal of Environmental Engineering* 122 (4).

U.S. EPA. 1994. Composting yard trimmings and municipal solid waste, EPA 530-R-94-003.

U.S. EPA. 2003. MSW characterization report—2003 data online document, http://www.epa.gov/epaoswer/non-hw/muncpl/pubs/03data.pdf. Accessed February 2008.

U.S. EPA. 2006. MSW characterization report—2006 MSW characterization data tables. Online document, http://www.epa.gov/epaoswer/non-hw/muncpl/pubs/06data.pdf. Accessed February 2008.

U.S. EPA. 2007. Municipal solid waste generation, recycling, and disposal in the United States: Facts and figures for 2006. EPA-530-F-07-030.

U.S. EPA. 2006. Municipal solid waste in the United States: 2005 facts and figures. EPA-530-R-06-011.

U.S. EPA. 2006. Municipal solid waste generation, recycling, and disposal in the United States: Facts and figures for 2005. EPA-530-F-06-039.

U.S. EPA. 2005. Municipal solid waste generation, recycling, and disposal in the United States: Facts and figures for 2003. EPA-530-F-05-003.

U.S. EPA 2007, http://www.epa.gov/epaoswer/non-hw/composting/index.htm, Last updated on Friday, September 7, 2007.

U.S. EPA. 2006. Landfill bioreactor performance, second interim report, Outer Loop Recycling and Disposal Facility Louisville, Kentucky. EPA/600-R-07-060.

U.S. EPA 2002a. State of the practice for bioreactor landfills. U.S. EPA 625/R-01/012; U.S. EPA: Cincinnati, Ohio.

U.S. EPA 2006a. www.epa.gov/reg3wcmd/solidwastesummary.htm. Accessed June 30, 2006.

U.S. EPA 2006b. http://www.epa.gov/eogapti1/module2/review2.htm. Accessed June 28, 2006.

U.S. EPA 2008. http://www.epa.gov/lmop/benefits.htm. Accessed, January 12, 2008.

Vogtlander, J. G., H. C. Brezet, and C. E. Hendricks. 2001. Allocation in recycling systems—An integrated model for the analyses of environmental impact and market value. *International Journal of Life Cycle Analysis* 6 (6): 344–355.

Wang, F., F. Roddick, A. Richardson, and R. Curnow. 1994. SWIM: Interactive software for continuous improvement of solid waste management. *Journal of Solid Waste Technology and Management* 22 (2).

Zero Waste Alliance, http://www.zerowaste.org/index.htm.

Zhuang, Y., S.-W Wu, Y.-L Wang, W.-X. Wu, and Y.-X. Chen. 2008. Source separation of household waste: A case study in China. *Waste Management*. 28 (10).

CHAPTER 6

HAZARDOUS WASTE TREATMENT

Mujde Erten-Unal
Old Dominion University
Norfolk, Virginia

1 INTRODUCTION AND DEFINITION OF HAZARDOUS WASTE

This chapter provides descriptive information on different hazardous waste treatment technologies. Different manufacturing and industrial processes generate hazardous waste. In addition, manufactured products are consumed throughout the society and lead to generation of hazardous waste by commercial, agricultural, institutional, and homeowner activities.

Hazardous waste is defined as any waste (solid, liquids, and containerized gases) listed in the EPA regulations that meets one of the characteristic of ignitability, corrosivity, reactivity, or toxicity and is declared as hazardous by the generator. This definition excludes the waste that is directly discharged into the air or water; these wastes are regulated under air and water laws that predate hazardous waste legislation (LaGrega et al. 2001). EPA regulates hazardous waste as a subset of solid waste.

The emergence of the Resource Conservation and Recovery Act in 1976 and its amendments in 1984 led to state-of-the-art technologies for the treatment of hazardous waste. The main purpose was to clean up air, water, land, and groundwater. The different hazardous waste technologies covered in this chapter are grouped under physicochemical treatment, biological treatment, thermal treatment, and land treatment categories.

2 PHYSICAL/CHEMICAL TREATMENT SYSTEMS

2.1 Stripping

2.1.1 Air Stripping

Air stripping removes volatile contaminants (VOCs), dissolved gasses, and semivolatile organic contaminants from the waste stream. It is mostly suitable for removing wastewaters with VOC concentrations less than 200 mg/l (LaGrega et al. 2001). VOCs are chemical compounds that have a tendency to evaporate faster. Some examples of VOCs can be treated with air stripping, including benzene, toluene, ethylbenzene, xylene (BTEX), and solvents such as trichloroethylene (TCE), dichloroethylene, chlorobenzene, and vinyl chloride.

Air stripping takes place in a stripping tower. The tower contains a packing medium. The wastewater contaminated with VOCs is introduced from the top of a packing media located in the tower. Air is introduced from the bottom of the tower. The contaminated water flows by forming a thin film over the medium. Stripping occurs when the wastewater and the air stream contact with each other and the dissolved molecules are transferred from a liquid into flowing gas. The governing equilibrium between the dissolved water phase and a gas phase (air) is based on Henry's law. According to Henry's law, partial pressure of a gas in the air above a dilute aqueous solution is proportional to its concentration in the solution. If Henry's law constant is high, the contaminant has low solubility in water and can be stripped easily. In addition, there is a direct correlation between Henry's constant and temperature: As temperature increases, air stripping efficiency increases as well.

Upon contact, the VOCs are vaporized and are collected in the air stream leaving the tower. The air stream exiting the tower needs to be further treated by using processes such as activated carbon. The different types of stripping towers include tray towers, packed towers, spray towers, diffused towers, and mechanical aeration towers.

2.1.2 Steam Stripping

Steam stripping is a process used to remove higher concentrations, up to several percent by weight of volatile organic compounds (VOCs), in an aqueous waste stream. It is also used to treat compounds that are not readily air-strippable, such as acetone, methanol, and pentachlorophenol. Nonaqueous wastes (such as spent solvents contaminated with nonvolatile impurities) are also good candidates for steam stripping. It has been successfully used for decontamination of groundwater containing ketones, alcohols, and chlorinated solvents (U.S. EPA 1987; Woodside 1999).

This process can be used in the treatment of industrial wastewaters, especially where the organic contaminant can be recovered for reuse or concentrated for more efficient destruction on site. Examples include recovery of solvents from an aqueous raffinate stream for recycling to an extraction process, or stripping of methanol, turpentine, and phenolic compounds from paper mill wastewaters. Due to high capital and operating costs, steam stripping is not likely to be considered for groundwater decontamination projects where air stripping is adequate.

Before treating a hazardous waste stream in a steam stripping process, suspended solids should be removed to avoid fouling the packing or plugging tray towers. Iron should be precipitated in a pretreatment step to eliminate oxidation in high temperatures that would create precipitation as ferric hydroxide ($Fe(OH)_3$).

There are two parts in the steam-stripping process: The bottom portion of the reactor works similar to an air stripper and the top portion enriches organic content of the steam to a point where a separate phase formation occurs. The liquid stream is saturated with the organic contaminants and sent back to the enrichment reactor. Important design considerations are the strippability of the organics and the capability of organics forming a separate organic phase in the overhead decanter (LaGrega et al. 2000). Any priority pollutant that is analyzed by direct injection to a gas chromatography and many compounds with boiling points less than 150°C are good candidates for steam stripping.

2.2 Activated Carbon Adsorption

Adsorption is a process of collecting soluble contaminants (the adsorbate) that are in solution on a solid surface (the adsorbent). It is used in removing organic contaminants from the waste stream. It is particularly useful when the chemical composition of the waste stream is not known. The most widely used adsorbent is activated carbon because it has excellent adsorption capacity for many different undesirable substances. It is made from wood, petroleum, and bituminous coal. Activation of carbon produces the porous structure that is essential for effective adsorption. This results in activated carbon having a very high surface area, ranging from 1,000 to 1,400 m^2/g (Watts 1997).

Activated carbon treatment had been identified by the U.S. EPA as the "best available technology" for the removal of most toxic organic substances (Federal Register 1987). Some of the compounds that can be removed by activated carbon include aromatic solvents such as benzene, toluene, xylene; polynuclear aromatics such as naphthalene and biphenyls; chlorinated aromatic compounds such as chlorobenzene, polychlorinated biphenyls (PCBs), aldrin, endrin, toxaphene, and

DDT; and phenolic compounds, soluble organic dyes, and chlorinated solvents such as carbon tetrachloride; and fuels such as gasoline, kerosene and oil.

Activated carbon can be used in the form of granular activated carbon (GAC) or powdered activated carbon (PAC). GAC has been used more commonly for the treatment of hazardous wastes because it is more effective in removing pollutants that have poor adsorption characteristics. It can also be regenerated by thermal methods and is therefore more economical. Powdered activated carbon (PAC) is more commonly used in biological treatment systems.

In theory, adsorption takes place at the surface of the activated carbon when the pollutant moves from the aqueous dilute wastewater into the carbon. The forces mainly responsible for adsorption onto activated carbon are the weak Van der Waals forces, electrical attraction, and the hydrophobic character of the organic material (LaGrega et al. 2001). The majority of organic molecules are adsorbed onto the large surface area within the pores of a carbon particle. The transfer of contaminants (adsorbate) from solution to the carbon surface (adsorbent) continues until equilibrium is reached with the concentration of contaminant remaining in solution.

The main factors influencing adsorption include magnitude and surface area of the adsorbent; pore size distribution of the adsorbent; nature and the concentration of the solute; temperature and pH of the solution; and the design and mode of operation of the activated carbon system. The equilibrium distribution of the contaminant between the liquid and solid phases helps define the capacity of a particular system. Therefore, adsorption isotherms are used to relate the concentrations of the adsorbed compound in each of the two phases. Isotherms are equilibrium relationships that would determine the degree to which adsorption will occur and the distribution of the contaminant between the liquid phase and the adsorbed phase at a specific temperature. For example, in the adsorption of a hazardous compound such as phenol, the isotherms determine what solid-phase phenol concentration in mg/gram corresponds to any given solution-phase concentration in mg/L.

To assess the feasibility of adsorption for a particular application, it is necessary to perform an adsorption isotherm study on the waste stream of interest. Data derived from an isotherm study will describe the performance of the activated carbon and will give important information. Some of the more familiar isotherm models are Langmuir and Freundlich.

Langmuir's isotherm is based on the assumption that all the adsorption sites have equal affinities for molecules of the adsorbate and that the presence of adsorbed molecules at one site will not affect the adsorption of molecules at an adjacent site. When a single contaminant is involved, Langmuir isotherms may provide closer approximations.

The Langmuir equation (Langmuir 1918) is commonly written as follows (Eckenfelder 2000):

$$x/m = abC/(1 + aC)$$

where:

x = amount of material adsorbed (mg or g)
m = mass of adsorbent (mg or g)
C = concentration of material remaining in solution after adsorption is complete (mg/L)
a and b = constants

If adsorption follows the Langmuir isotherm, a linear relationship should result when the inverse of x/m is plotted against the inverse of C. Values of the constants a and b can be determined from the slope and intercept of the plot.

Freundlich (1926) developed an empirical equation to describe the adsorption process (Eckenfelder, 2000). His development was based on the assumption that the adsorbent had a heterogeneous surface composed of different classes of adsorption sites, with adsorption on each class of site following the Langmuir isotherm. Freundlich offered the following equation:

$$x/m = KC^{1/n}$$

where:

x = amount of solute adsorbed (mg, g)
m = mass of adsorbent (mg, g)
C = concentration of contaminant remaining in solution after temperature
K and n = constants that must be evaluated for each solute and temperature

The Freundlich equation can be put in a useful form by taking the log of both sides. If the plot of log x/m versus C yields a straight line, then the adsorption data should follow the Freundlich theory.

In a granular activated carbon process, a continuous-flow fixed-bed column is often used as a way of contacting wastewaters with GAC. The wastewater is applied to the top of the column and withdrawn from the bottom. The carbon is held in place with an underdrain system at the bottom of the column. In an attempt to limit the headloss build-up due to the removal of particulate matter, backwashing and surface washing are usually required within the carbon column. A GAC system must allow for the removal of spent carbon-regeneration and addition of new carbon.

A laboratory absorption study should be conducted to evaluate the feasibility and economics of adsorption. Column tests simulate the actual operation of a full-scale unit. In the laboratory, 2-inch-diameter columns are filled with the carbon to be tested and the contaminated wastewater runs through. The volume within the carbon bed in which adsorption occurs is called the *mass transfer zone* (MTZ, or adsorption zone). After a certain volume of the wastewater passes through the carbon column, breakthrough is reached, which indicates the point that influent contaminant concentration is detected in the effluent coming off the carbon column (Eckenfelder 2000). Contaminant concentration in the effluent will increase after breakthrough. However, if one runs the carbon columns in the series, the carbon bed in the first column is fully utilized before reaching the second column, and not much of the contaminant will be released to the environment.

The major design considerations include flow rate and headloss, nature and amount of organic compounds present, empty bed contact time required, filtration rate, and carbon regeneration frequency and methods. It is feasible to regenerate spent carbon for economic reasons. Different modes of regeneration include thermal methods, solvent extraction, acid or base treatment, and chemical oxidation. When determining the suitability of using GAC for specific hazardous waste, the change in capacity of carbon through consecutive regeneration cycles should be taken into consideration (Eckenfelder 2000).

2.3 Oxidation

2.3.1 Chemical Oxidation

The objective of chemical oxidation is to detoxify waste by adding an oxidizing agent and to chemically transform waste components to compounds such as carbon dioxide and water. It is a well-established technology capable of treating a wide range of liquid hazardous waste that include organic compounds such as pesticides, phenols, detergents, chlorinated VOCs, phenolic waste, wastes with low organic content, and cyanide. The chemicals that are reduced are the contaminants.

Oxidation-reduction reactions occur in pairs to form an overall REDOX reaction. Oxidizing agents are nonspecific and will react with any reducing agents present in the waste stream.

Hazardous waste treatment by chemical oxidation involves mixing two liquid streams—the waste and the treatment chemical—or contacting the aqueous solution with gas. The mixing reactors can be in batch, completely mixed, or plug flow mode. Mixing can be provided by mechanical agitation, introducing air through the reactor.

The oxidizing chemicals have the following properties:

- *Ozone*. Ozone is a very powerful oxidant. It is unstable and dissociates into side reactions rapidly. Its high free energy indicates that the oxidation reaction may proceed to completion. Ozone dissociates to oxygen very rapidly and must be generated on site.
- *Hydrogen peroxide*. Hydrogen peroxide is effective in oxidizing toxic hazardous wastes and cyanide-bearing wastes. Hydrogen peroxide generates hydroxyl radicals (·OH) in the presence of a catalyst such as iron. The radical reacts with organics and reduced compounds to produce a reactive organic radical (·R). The organic radical reacts again with peroxide to

produce additional hydroxyl radical, which, in turn, oxidizes pollutants to more readily biodegradable compounds.

- *Chlorine*. Chlorine and its various compounds are used extensively in water and wastewater treatment. Chlorine is the principal chemical involved in disinfection. However, when combined with organic material, chlorine forms trihalomethane (THM), which is carcinogenic. Chlorine is evaporated to a gas and mixed with water to provide a hypochlorous acid (HOCl) solution. Hypochlorous acid is then converted to hypochlorite ion. The reaction is pH dependent and as pH increases the oxidation power increases as well. One of the most commonly used hazardous waste treatment process is oxidation of cyanide by alkaline chlorination process. The purpose is to add chlorine and convert cyanide to cyanate. However, pH must be greater than 10 to prevent the formation of cyanogen chloride, which is a very toxic gas and can exist at lower pH levels. Cyanate produced is less toxic and hydrolyze under acidic conditions in lakes and streams to ammonia and carbon dioxide. In this reaction, a pH of 8.5 is used to simplify operation to permit the last two reactions to occur sequentially in a single treatment unit (Weber 1972).

2.3.2 Supercritical Water Oxidation

Supercritical water oxidation (SCWO) is an emerging technology in which dilute concentrations of organic and inorganic wastewaters are oxidized under high temperature and pressure. The temperature and pressure conditions of the water are elevated to supercritical conditions that exert properties between those of gas and liquid. When air and contaminated water are brought together above the critical point of water, complete oxidation of organic contaminants occur rapidly. As a result, organic contaminants dissolve and become soluble in the high temperature and pressure environment exerted by the supercritical water, and then they become oxidized.

Major design considerations of SCWO include the reactor and heat exchanger. Design involves considerations of residence time, temperature, and materials of construction. Precipitated inorganic salts may adhere to the reactor walls and decrease effective volume available for the reaction and reduce residence time in the reactor. In addition, the reactors must be resistant to corrosion because they are exposed to high temperatures and high pressure. In the SCWO treatment, one advantage is that organic compounds are destroyed rather than being removed, and the chemical reactions are carried out in a closed system where there is physical control and maintenance. The process is capable of generating all power required for air compression and feed pumping.

Supercritical water oxidation has been proven to be effective technology for the removal of organic compounds such as trichloroethylene (TCE), trichloroethane (TCA), PCBs, organochlorine insecticides, dyes, pulp and paper wastes, and chlorinated dioxins (Jensen 1994; Woodside, 1999; Watts 1997).

2.3.3 Catalytic Wet Air Oxidation (CWAO)

Wet oxidation can be defined as the oxidation of organic and inorganic substances in an aqueous solution or suspension by means of an oxidant (usually oxygen or air, but sometimes ozone and hydrogen peroxide) at elevated temperatures and pressures. Typical conditions for wet oxidation range from 180°C and 2 MPa to 315°C and 15 MPa. Residence times may vary from 15 to 120 minutes. Insoluble organic matter is converted to simpler organic compounds, which are oxidized and eventually converted to carbon dioxide and water, without emissions of NO_x, SO_2, HCl, dioxins, furans, or ash. The last residual organic compounds are carboxylic acids, especially acetic acid (Luck 1999). Catalytic wet oxidation employs catalysts to reduce the severity of the already-mentioned reaction conditions. Compared to conventional wet air oxidation, catalytic wet air oxidation offers lower energy requirements and much higher oxidation efficiencies. Current catalytic wet oxidation processes rely either on supported precious metals and/or base metal oxide catalysts, or on homogeneous catalysts such as Fe or Cu (Pintar 2003). Matatov-Meytal and Sheintuch (1998) state that a catalyst for aqueous phase oxidation should have high oxidation rates, should be nonselective, should be physically and chemically stable in hot acidic solutions and mechanically strong and resistant to attrition, and should maintain a high activity for a prolonged use at high temperatures.

2.4 Membrane Filtration—Reverse Osmosis

Membrane filtration involves separation of the contaminant (solute) from a liquid phase (solvent), typically water. The objectives of the process are reduction in volume of waste, recovery or purification of liquid waste, and concentration and/or recovery of the contaminant. Membranes used to retain material are based on molecular size and shape. Most common membrane filtration processes include reverse osmosis, electrodialysis, and ultrafiltration.

Osmosis is a process in which water moves under osmotic pressure to establish equilibrium in the ionic strength of solutions across a semipermeable membrane. A thin membrane separates waters with different salt concentrations. The water moves from the more dilute side of the membrane to the more concentrated side. In reverse osmosis, water is forced through the semipermeable membrane from the concentrated side to the dilute side against natural osmotic pressure, as shown in Figure 6.1. The driving force for mass transfer process is the pressure gradient. The rate of flow through reverse osmosis membrane is directly proportional to effective pressure.

The different kinds of membranes include cellulose acetate membranes, which have high flow rate per unit area. They are used as spiral modules, which are made of large membrane sheets. Another type is polyamide membranes, which have a lower specific flow rate and are manufactured to form hollow fibers in pressure vessel.

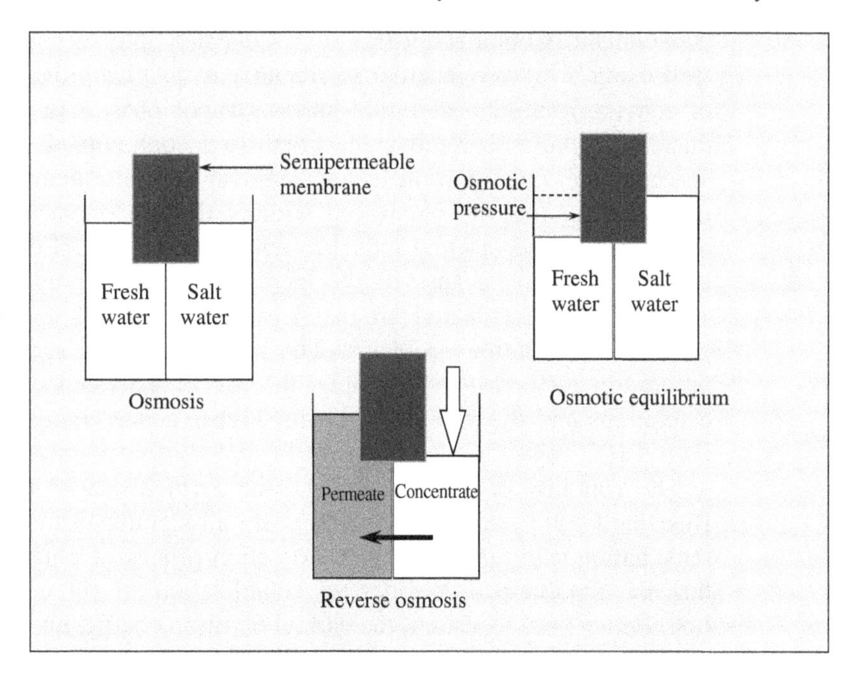

Figure 6.1 Description of reverse osmosis process

An important design parameter is flux, which is the volume of water that can flow through a unit area of membrane. Flux is a function of pressure differential between applied pressure and osmotic pressure across the membrane.

The membranes are subject to fouling. In hazardous waste management, membranes are limited to extremely toxic materials that cannot be removed by cost-effective technologies. Reverse osmosis is used to remove metals from wastewater, remove dyes from textile industry, and recover oil in emulsified form.

2.5 Soil Vapor Extraction

Soil vapor extraction is also known as vacuum extraction, in situ volatilization, soil venting, in situ aeration (Wong and Nolen 1997), and soil vapor stripping (Riser-Roberts 1998). It is commonly used because it is both a low-cost and effective hazardous waste remediation method (Wong and Nolen 1997).

The main idea of SVE is to reduce the vapor pressure within the soil to enhance volatilization of the petroleum hydrocarbons so that they can be removed by vacuum extraction (Riser-Roberts 1998). Petroleum hydrocarbons in the unsaturated zone of the soil may be in several forms: "vapor phase, dissolved phase (in pore water), liquid phase or nonaqueous phase liquids, and adsorbed phase" (Wong 1997). Injection and extraction are both utilized to remove the volatile organic contaminant. Contaminant-free air is injected into the soil through a vertical well using a blower in order to implement volatilization, which is the process of converting a solid or liquid into the vapor phase. An extraction well is put under a vacuum to remove the volatilized contaminant from within the pore space of the soil. In some cases, removal of the contaminated air is passive, meaning that the air is removed without the use of vacuum pressure through a perforated pipe inserted into the contaminated area. However, active removal is more efficient because of the increased pressure gradient (Chambers et al. 1991). The contaminated air that has been extracted must then be treated to prevent air pollution problems. The gas stream is normally treated using an adsorption method—for example, the use of activated carbon for lower concentrations and thermal oxidation for higher concentrations of the contaminant (U.S. EPA 2006).

The volatility of the contaminant, the permeability of the soil (Stamnes and Blanchard 1997), and site conditions are all significant factors in the success of remediation using the soil vapor extraction process. Contaminant characteristics that are important to consider for volatilization are the contaminant concentration, Henry's law constant, the carbon partition coefficient, vapor pressure, water solubility, melting point, and the boiling point (LaGrega et al. 2001; Wong 1997). Soil vapor extraction is found to be most successful for volatile organic compounds, with a Henry's constant of 0.01 or a vapor pressure of 66 Pa or greater (DePaoli et al. 1996). Soil properties of specific sites must be taken into consideration, because it is essential for the flux of air to move through the contaminated soil. For this reason, the air permeability and moisture content of the soil must also be considered (Wong 1997).

Once soil vapor extraction has been selected as the method of treatment, it is necessary to design a system that will be efficient in cleaning the site and that will be reasonable in cost and duration. Modeling and pilot studies are important in the process of designing a SVE system specific to a contaminated site that will remove contaminants effectively.

SVE is a successful method for the remediation of petroleum-contaminated soils. However, the process is limited to contaminated soils that are unsaturated and permeable. Examples of suitable soils are gravels, sands, and coarse silts. The system is very flexible in design, which makes it a more useful method of remediation because it can operate specifically for a certain site. These parameters include the spacing of the extraction wells, rate of air injected into the soil, and variations in pumping. More wells can be added if necessary, but more wells added to the system will increase the cost. Also, the more time the injected air spends in the contaminated soil, the more efficient the treatment process will be because it will allow plenty of time for the contaminant to vaporize. Higher amounts of the contaminant can also be removed by increasing venting (Chambers et al. 1991).

3 BIOLOGICAL TREATMENT SYSTEMS

Biological treatment involves the application of microorganisms that utilize carbon and energy for growth to oxidize the organic matter present in the waste stream through biochemical means under controlled conditions. The objective of biological treatment is to promote and maintain microbial population (biomass) that metabolizes (biodegrade) a target waste. The waste material that is adsorbed by microorganisms is subject to different biochemical reactions, which take place at temperatures ranging between 0°C and 40°C.

In order to carry out biochemical reactions at low temperatures, catalysts must be present to lower the activation energy of these reactions. Catalysts are present as enzymes, which are organic compounds produced by living organisms in their life process. There are intracellular enzymes that carry out biochemical reactions inside the cell and extracellular enzymes that are excreted from the cell to carry out reactions outside the cell. In biological degradation, large organic molecules are broken up by extra cellular enzymes through *hydrolysis*. They are then subject to intracellular enzymatic reactions. Enzymes' role in metabolism is to lower the energy required to activate a reaction and thereby speed up biological activity.

Microorganisms use catabolic and anabolic metabolism. Catabolic metabolism releases energy that is captured and transformed to support cell maintenance and cell building activities. Energy release involves the transfer of electrons from organic carbon. To complete this reaction, electron acceptors such as oxygen, nitrate, and sulfate are needed. In catabolic metabolism, oxygen is consumed and organic matter is removed. Anabolic metabolism is a cell-building metabolism. The anabolic processes produce protoplasm as an end product, which is composed of proteins, carbohydrates, DNA, and other components (LaGrega et al. 1994).

The biological degradation of an organic compound involves the transfer of electrons from the waste to an electron acceptor. This process is called *respiration*. There are two types of respiration: aerobic and anaerobic. In aerobic respiration, bacteria use oxygen as a terminal electron acceptor. Here, oxygen is reduced to water and organic carbon is oxidized to carbon dioxide. In anaerobic respiration, terminal electron acceptors can be of inorganic compounds such as nitrates being reduced to nitrogen gas; sulfates being reduced to hydrogen sulfide gas; or carbon dioxide being reduced to methane.

During the biodegradation process, microorganisms must have the optimum environmental conditions for cell growth. These are optimum temperature, pH, moisture, and macro and micro nutrients.

Biodegradation of hazardous wastes can be accomplished using suspended-growth or attached-growth biological systems. Suspended-growth treatment systems keep the microorganisms suspended freely in water. The microorganisms are maintained in suspension within the liquid, and microorganisms convert the organic matter or other constituents in the wastewater into gases and cell tissue. The most common type of aerobic system is the suspended-growth treatment system. Suspended-growth technologies are conventional activated sludge-treatment systems that use various process modes, sequencing batch reactors, and anaerobic treatment reactors.

In attached-growth biological systems, the microorganisms are attached to a supporting media and form a biofilm on the surface of the media. The media provide a large surface area for attached growth. The media also immobilize and retain the biomass in the reactor so the solid retention times can be longer than hydraulic retention times. The media consist of crushed stones, plastic, tile, or slag. After a prolonged time period, microorganisms in the biofilm become old and die. The dead microorganisms can no longer hold on to the media, and the biofilm detaches itself from the media. This is called *sloughing*. The packing of the media may be packed (down-flow) or fluidized (up-flow). The most common types of attached-growth systems are trickling filters and rotating biological contactors.

3.1 Activated Sludge

An activated sludge is a suspended-growth system in which a pretreated wastewater is aerated to promote the growth of microorganisms (biomass) that slowly consume the organic compounds in the wastewater. The microorganisms acclimate to the specific mix of compounds present in the wastewater and therefore use up the organic portion significantly. The activated-sludge process consists of a reactor in which the microorganisms responsible for treatment are kept in suspension by mixing and aeration; and a final clarifier where liquid–solids separation occurs. Biomass that is captured in the clarifier is recycled back to the reactor, depending on the configuration of the reactor. Contact time between waste and biomass is also controlled by wasting excess biomass. Equalization is necessary for wastes that have variable characteristics to eliminate the inhibition of bacteria. Equalization can be achieved by using storage tanks before pumping the waste to the biological reactor.

Floc formation in the aeration tank is a very important feature of the activated sludge process. These flocs will form flocculent settable solids that can be removed by gravity settling in the final clarifier (Eckenfelder and Musterman 1995). The clarification allows for a clear effluent, as well as serves an additional microorganism source for the process. The process diagram of activated sludge is shown in Figure 6.2.

Activated sludge is commonly applied to municipal and industrial wastewater treatment, with several variations of the two basic process configurations. The selection of the "best" process depends on the waste characteristics. What follows are the process design considerations.

Activated sludge exists in two general operating modes: (1) complete mix with or without solids recirculation or (2) plug flow with or without solids

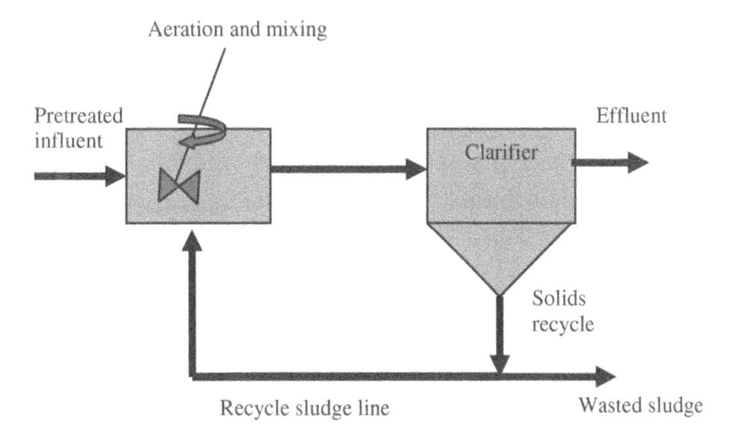

Figure 6.2 Activated sludge process

recirculation. In a complete mix reactor, the influent is dispersed immediately upon introduction, so there is no concentration gradient in the tank. The reactor contents are completely mixed, and it is assumed that there are no microorganisms in the influent. Uniform concentration of the contaminants is maintained throughout the reactor. In a plug-flow reactor, there is no longitudinal mixing, and all particles entering the reactor stay in the reactor an equal amount of time. Some particles may make more passes through the reactor because of recycle but while they are in the tank, they all pass in the same amount of time.

Solid retention time (SRT) is the average time that the microorganisms spend in the reactor. It is the ratio of mass of organisms in the reactor to the mass of organisms removed from the reactor. In the design of activated sludge systems, SRT is one of the most critical design parameter (Metcalf and Eddy 2002). An increase in SRT will increase concentration of biomass in the reactor. If SRT is too short, then the system is removing too many microorganisms and biomass will wash out. SRT can be controlled by wasting more or less sludge from the recycle line.

Hydraulic retention time (HRT) is the average amount of time that the wastewater stays in the reactor. HRT is determined by dividing the reactor volume by the wastewater flow rate. SRT is greater than HRT, and it is independent of HRT. In successful operations, a minimum HRT must be met before SRT becomes a controlling factor (Metcalf and Eddy 2002).

Food to microorganism ratio shows the organic load entering the reactor. It is used as one of the operational design parameters. The ratio is calculated by dividing the BOD load with the bacterial mass or biomass load expressed as mixed liquor suspended solid or mixed liquor volatile suspended solids in the reactor. It keeps the nutrients balance in the reactor and controls the overgrowth of filamentous microorganisms to avoid unsettling sludge in the secondary clarifier.

Aeration is required by the microorganisms for the stabilization and degradation of the waste. It is recommended that the aeration equipment be designed with a safety factor of at least two times the average BOD load. Aeration equipment should be sized to leave residual dissolved oxygen of 2 mg/L at average load and 0.5 mg/L at peak load. The air supply system should be capable of providing oxygen to meet the diurnal peak oxygen demand or 200 percent of the design averages whichever is greater.

3.2 Sequencing Batch Reactors (SBR)

Sequencing batch reactors are fill-and-draw suspended growth batch treatment systems where the waste treatment takes place in the same reactor at different time sequences. At each specific time period, a specific operation takes place. The operational steps are: idle, fill, reach, settle, and draw (U.S. EPA 1999). The idle step takes place between the end and beginning of the operational cycle. During the fill step, influent wastewater is added to the reactor. The fill step can be aerated, mixed, or kept static. During the react stage, the reactor contents are either aerated or mixed depending on the operational mode. After the react stage, the biomass is allowed to settle for a certain period of time, and the supernatant is decanted during the draw stage. SBRs are useful in the treatment of wastes having high concentrations of organic compounds as high as 15,000 to 20,000 mg/l of chemical oxygen demand (COD), such as in the case of landfill leachate wastewaters. There are several advantages to using SBRs for leachate stream treatment, including the traditional SBR benefits: process flexibility, nearly ideal quiescent settling, no external clarifiers required, and elimination of short-circuiting. The process flexibility of the SBR process is invaluable for highly variable waste streams, which is why it is often used in industrial processes.

3.3 Aerated Lagoons and Facultative Lagoons

Aerated lagoons and facultative lagoons are suspended-growth basins in which wastewater is treated on a flow-through basis. In an aerated lagoon, oxygen is supplied by surface aerators or diffused air units, and there is photosynthetic activity throughout the entire water column. Aerated lagoons are usually followed by a settling pond to allow removal of pathogenic microorganisms.

In facultative lagoons, treatment is achieved through bacterial action in all layers (all aerobic, anaerobic, and anoxic). Facultative lagoons range 3 to 8 feet in depth with an anaerobic lower layer, facultative middle zone and aerobic upper zone. Oxygen enters through wind-induced mixing and aeration naturally. The design is based on long enough detention time and low organic loading rate to achieve aerobic conditions on the surface. Settled solids are digested in the anaerobic zones.

3.4 Anaerobic Treatment

Anaerobic treatment is a biological process carried out in the absence of oxygen for the stabilization of organic materials by conversion to methane (CH_4) and inorganic end products such as carbon dioxide (CO_2) and ammonia (NH_3). Anaerobic conversion occurs in three steps:

- *Hydrolysis*. In hydrolysis, water directly participates and breaks the chemical bond within the molecules. Complex organic materials such as protein, carbohydrates, and lipids are converted to simple compounds such as amino acids, sugars and long-chain fatty acids, respectively, by the action of enzymes excreted by fermentative bacteria.
- The next step involves bacterial conversion of high-molecular-weight compounds into low-molecular-weight intermediate compounds for use as a source of energy and cell tissue.
- The final step involves conversion of intermediate compounds into simpler end products such as methane and carbon dioxide by methane-generating bacteria.

The successful operation of an anaerobic reactor depends on maintaining the environmental factors close to the comfort of the microorganisms involved in the process. Since methane generation is the rate-limiting step in anaerobic treatment of wastewater, the major environmental factors are governed by the methane-producing bacteria. Some of the important environmental factors are temperature, pH, availability of nutrients, presence of sufficient alkalinity, and absence of toxic and inhibitory compounds in the influent.

Best candidates of industrial wastewaters for anaerobic treatment are petrochemical wastes, pulp and paper wastes, alcohol production, brewery waste, and chemical wastes (LaGrega et al. 2001). Low-rate and high-rate anaerobic treatment processes are used to treat hazardous wastes. High-rate anaerobic systems are able to retain very high concentrations of active biomass in the reactor. Therefore, these systems can maintain very high SRT levels. Some of the high-rate anaerobic treatment processes are anaerobic contact process, anaerobic filter, up-flow anaerobic filter, down-flow anaerobic filter, and up-flow anaerobic sludge blanket.

3.5 Trickling Filters

Trickling filters are attached-growth treatment systems where wastewater passes over a fixed bed of packed media covered with biofilm. The media are either rock or plastic, and the wastewater flows over the media where bacteria grow and creates the biofilm or biological slime layer. As the wastewater flows over the biofilm, organic matter (substrate), oxygen, and nutrients diffuse across the boundary layer. Organic matter removal is a function of the available biological slime surface and the time of contact of the wastewater with that surface. The reaction rate is limited by how much material diffuses through. Within the biofilm, substrate is utilized for biological growth. Biofilm loses mass continuously due to erosion of small pieces or sloughing of large pieces. Wastewater is dispersed with rotating spray distributors that spin due to the jet action of sprays. They are simple to operate, resistant to shock loadings, and require lower power. However, the BOD removal rates are lower than activated sludge systems and they may have higher suspended solids in the effluent. The two types of operating trickling filters are low-rate and high-rate trickling filters.

3.6 Rotating Biological Contactors

A rotating biological contactor (RBC) is an attached-growth biological treatment system that contains approximately 10- to 12-foot-diameter closely spaced plastic numerous discs that are connected with flow across shaft located in a rectangular basin. The rectangular basin is filled with the wastewater that needs to be treated. Approximately less than half of the discs are immersed in wastewater. As the discs rotate at a slow speed, they submerge into the wastewater and aerate as they come out of the wastewater. Biofilm continuously grows on discs as the rotation brings microorganisms in close contact with the wastewater. Rotation also causes shearing action and causes sloughing of biofilm, keeping a constant biofilm thickness. Aeration is provided by the oxygen transfer into the biofilm while the discs are rotating outside of wastewater and by mixing turbulence that entrains air into tank liquid. The discs are generally covered for temperature control and to prevent photodegradation. The RBCs are used to treat low-strength industrial wastewaters due to insufficient oxygenation provided by this system (Woodside 1999).

3.7 Bioremediation

Bioremediation is the use of biological organisms to aid in the removal of hazardous waste from a contaminated area. The goal of bioremediation is to stimulate microorganisms with nutrients and other chemicals that will enable them to destroy the contaminants. Bacterial bioremediation becomes an engineered process when the natural bacterial feeding, growth, and reproduction cycles are enhanced in some way as to increase the rate at which contaminants are removed from a hazardous waste site. The method for enhancing these cycles is by optimizing the environmental conditions for the bacteria, whether through increased nutrient loading; stabilizing and/or maintaining ideal temperature, light, oxygen, or moisture conditions; or other parameters that are beneficial for a bacteriological community.

In hazardous waste management, some contaminants are more difficult to treat than others. Some treatment processes, even though they work effectively, may be prohibitively expensive. Bioremediation presents itself in many cases as a less costly solution to these problems. It can break down some of the quite hazardous and persistent contaminants in the environment in a cost-effective way. Bacterial bioremediation is normally an *in situ* treatment method, meaning there are no significant transportation requirements, need for heavy equipment, or expensive cleanup procedures. Since this process is fundamentally natural, cost is nominal and the process may be able to proceed with a minimum of human interaction (Chapell 1995). As far as treatment is concerned, basically all of the uses of bioremediation have the same goal in mind—reduction of some unwanted contaminant—as well as the same technique in which the bacteria consume or degrade the contaminant to constituents that can easily be treated by other means. The types of microbial processes that will be employed in the cleanup dictate what nutritional supplements the bioremediation system must supply. The byproducts of microbial processes can provide indicators that the bioremediation is successful.

There are two types of bioremediation:

1. Intrinsic bioremediation. The native subsurface microbes degrade the contaminants without direct human intervention. This does not accelerate the treatment but prevents spread of further contaminants.
2. Engineered bioremediation. This is the acceleration of microbial activity using engineered site-modification procedures such as installation of wells to circulate fluids and nutrients to stimulate microbial growth. Air is often forced into the ground to supply needed oxygen for the bacteria. The process can be accelerated by increasing airflow and warming up the soil temperature using heat wells, which provide better conditions for microbial activity. The principal objective of engineered bioremediation is to isolate and control contaminated field sites so that they will become in situ bioreactors.

Some of the hazardous compounds that can be treated by using bioremediation are petroleum hydrocarbons and derivatives such as gasoline compounds (benzene, toluene, ethylbenzene, and xylene), fuel oil, polycyclic aromatic hydrocarbons (PAHs), creosote, ethers, alcohols, ketones and esters. Others are halogenated compounds with halogen atoms (usually chlorine, bromine, or fluorine) added to them in place of hydrogen atoms and nitro aromatics, which are organic chemicals, and with the nitro (-NO_2) bonded to one or more carbons in a benzene ring. An example is the chemical used in explosive laboratories trinitrotoluene (TNT). Although microorganisms cannot destroy metals, they can alter their reactivity, mobility, and toxicity.

3.8 Phytoremediation

Phytoremediation uses green plants to assist in cleaning up contamination from soils, groundwater and surface waters. Phytoremediation may be conducted near surfaces of soils, in situ in the deep aquifer, or ex situ for the treatment of contaminated liquids by extracting groundwater or surface water. The plants can excrete organics that can serve as additional substrates for the bacteria that degrade hazardous contaminants. The plant roots can serve as adsorption sites and can provide more oxygen into the soil. The plants can also uptake organics and degrade them by using their enzymes or can respire them into the air, called *phytovolatilization*, which helps remove compounds such as trichloroethylene (TCE), benzene, toluene, ethyl benzene, xylene (BTEX), and chlorinated benzenes (LaGrega et al. 2001).

Plants can uptake metals and transform them to less toxic forms by changing their redox states, such as in the case of conversion of toxic hexavalent chromium to trivalent chromium. Plants can also accumulate metals in their roots and above-ground portions and remove them from contaminated soils. Several applications of phytoremediation include in situ cleanup, where plants uptake contaminants from soil or groundwater; and ex situ cleanup, where plants are grown in contaminated water, which is also referred to as hydroponics.

The treatment efficiency is depends on the ability of plants' uptake of the constituents by the root system through the plant stems and leaves. Success also depends on the contaminant of concern, existing vegetation, and potential risk to humans.

4 THERMAL TREATMENT SYSTEMS—INCINERATION

Incineration is the process of controlled burning at high temperatures of solid, liquid, or gaseous wastes. It reduces the volume and weight of hazardous waste to a fraction of its original size. This process also converts hazardous organic compounds to ash. The three most important operating conditions for proper incineration are temperature, residence time, and turbulence, which are called the three Ts of incineration. Incineration of waste materials converts hydrogen to water vapor, chloride or fluoride to hydrochloric acid (HCl), or hydrofluoric acid (HF), carbon to carbon dioxide (CO_2), sulfur to sulfur dioxide (SO_2), alkali metals to hydroxides, and nonalkali metals to oxides. Waste consistency depends on the type of waste produced and can be categorized into the following:

- Sludge materials are too viscous, abrasive, or varying in consistency.
- Wastes undergo partial or complete phase change during incineration.
- High residue materials, high ash liquids, and sludge and materials contain salts or metals.

4.1 Gaseous Waste Incinerators

Incineration for gaseous wastes may be divided into three classifications: direct flame, thermal, and catalytic. *Direct-flame incineration* is typically applied where the gas stream has sufficient energy value to maintain a flame without the need to provide a supplemental fuel source. A common application of direct-flame incineration is the flare. Flares are used to burn landfill and digester gases, as well as waste refining gases.

Thermal incineration is applied to gas streams with low energy value that require supplemental fuel to carry on combustion. These gases may have heating values and operate at lower temperatures. If the incinerator is missing in any of the three Ts, carbon monoxide and intermediate combustion byproducts may be formed. However, under proper operating conditions, significant destruction efficiencies can be achieved.

Catalytic incinerators use catalysts to enhance the oxidation at desired temperatures. Due to the presence of a catalyst, the oxidation rate of wastes can be achieved at lower temperatures and shorter residence times. The catalyst is subject to deterioration at sudden increases in temperatures; therefore, the VOC content of the waste being fed to the incinerator should be kept homogenous.

4.2 Incineration of Liquid Wastes

For the incineration of liquid waste, proper mixing is necessary in order to feed uniform mixture of waste into the incinerator. If liquids do not vaporize at ambient temperatures, they must be injected into the furnace with an atomizer in the form of fine spray. This would help proper mixing of combustion air with the liquid droplets and will allow combustion to proceed. There are primary and secondary combustion units. The wastes with high heating values are burned in the primary combustion chambers, and remaining organics are oxidized in the secondary combustion chamber for complete oxidation.

4.3 Incineration of Solid Wastes

Solid waste is more difficult to incinerate and requires processing for size reduction prior to burning. Since most solid waste contains materials that are not completely combustible, there will also be residual ash. Incinerator types employed for solid wastes include the following:

Controlled air incinerators are incenerators with a two-chamber fixed heart system where wastes are vaporized under low air conditions in the first chamber and then ignited with auxiliary fuel in the second chamber. It is constructed in a modular form and has low capital cost.

Rotary kiln incinerators use a rotating horizontal cylinder that partially mixes the waste to improve combustion. They can be continuous or batch fed. Rotary kiln incineration works well with emission control equipment. It is used for wastes with higher solids content or larger solids size and has high ash residual.

Fluidized bed incinerators consist of a bed of sand or alumina in which the waste is fed continuously for upward flow through a fluidized bed. Air flow provides good mixing and keeps the particles in bed in suspension. This system has low oxygen requirements, lower air emissions, and is suitable for burning wastes with lower heating values. It also works well for wastes with smaller solid size and generates less ash than rotary kiln incinerators.

Multiple hearth incinerators have limited use in hazardous waste treatment because the temperatures cannot be raised high enough to destroy the hazardous constituents due to equipment limitations. They are more commonly used for sewage sludge treatment. Waste is fed from the top. As it combusts, residues are raked to the center, where they fall on progressively lower hearths until only ash remains. These are usually larger unit processes. They are used for wastes with higher solids content or larger solids size and generates higher ash residue.

5 LAND DISPOSAL SYSTEMS

One of the options for ultimately disposing of hazardous wastes is landfilling. Landfills are relatively inexpensive when compared to other treatment and disposal options, but they do have some negatives. U.S. federal regulations (RCRA) developed in the late 1980s require separate landfill disposal of municipal (household) solid waste and hazardous waste. RCRA Subtitle D regulates municipal solid waste (MSW) landfills, and RCRA Subtitle C is used for hazardous waste landfills. Before RCRA, most landfills were dumps. Many closed dumps still exist, and a majority of these dumps continue to pollute groundwater and surface waters.

Hazardous wastes that are disposed of in a landfill in many cases still exist in toxic form, but are suitably contained for landfill disposal. However, the toxicity will remain and has the potential to leak/leach into surrounding soils and groundwater. For this reason, the landfill liners and the leachate collection and treatment systems are viewed to be the two most important factors when designing a new landfill.

In weighing the options for hazardous waste disposal, landfills are often determined to be the best option, based primarily on disposal costs. A major portion of hazardous wastes are disposed of in this method. According to Visvanathan (1996), landfilling accounts for 79 percent of the hazardous waste disposal in the United Kingdom, storing 2.7 million tons per year of waste. A similar percentage is likely in the United States.

The most important landfill design considerations when designing a hazardous waste landfill include site selection, properly installing clay and synthetic liners to limit leachate percolating into the groundwater, the final cap to a closed landfill, and the leachate and gas collection and treatment systems.

Wastes have to be contained or stabilized prior to disposal in a landfill site. In order for the wastes to be stored in the landfill, they must be contained in sealed drums or other containers or other stabilization to limit the waste's migration from its landfill cell into other areas of the landfill, where there may be leachate infiltration into the groundwater. Leachate is the liquid that forms at the bottom of the landfill, resulting from percolation, precipitation, uncontrolled runoff, and

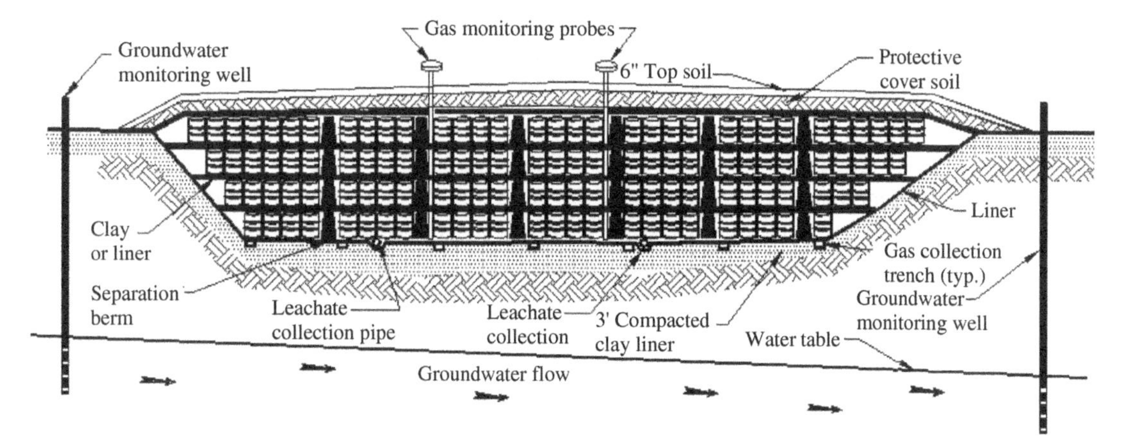

Figure 6.3 Schematic of a hazardous waste landfill with drums

irrigation water into the landfill. A schematic of a hazardous waste landfill is shown in Figure 6.3.

According to Slack et al. (2005), the codisposal of hazardous and nonhazardous wastes is a practice soon to be banned in EU member states through the implementation of the landfill directive, effective from July 2004. This will potentially reduce the volume of leachate created in an atmosphere where there is mixed waste storage.

In a hazardous waste landfill, the primary purpose of liners is to minimize leakage of landfill leachate and gas into the subsurface and to allow collection of leachate for treatment and disposal. Liners consist of successive layers of compacted clay and/or geosynthetic material designed to prevent migration of landfill leachate and landfill gas. Most commonly used materials for liners include one or all of the following:

- Geomembrane (Hydraulic barrier)
- Geosynthetic clay liner (GCL) (hydraulic barrier)
- Compacted clay (hydraulic barrier)
- Geotextile (for cushion or separation)

To promote removal of leachate from the liner, the landfill bottom is sloped and sufficient drainage pipes are provided to ensure that the leachate depth over the liner does not exceed 1 feet. Typical slope of base liner is 2 to 10 percent and typical slope of sidewall liner ranges from 20 to 40 percent in a landfill (McBean et al. 1995).

The main function of a leachate collection system (LCS) is to collect leachate for treatment and disposal. Most commonly used materials for LCS include pea gravel or coarse sand, geocomposite drainage layer (GDL), perforated HDPE/PVC pipes, leachate sump, and a submersible pump. Due to the leakage potential of all types of liners, landfill leachate collection systems have multiple leachate collection zones to protect groundwater from contamination. The secondary leachate collection zone—also known as leak detection system—controls the leachate that may pass through the primary LCS (LaGrega et al. 2001).

Leachate contains many of the same constituents that are typical contaminants of concern when considering disposal of any waste stream, especially very high COD and ammonium concentrations. Unique to leachate, however, are xenobiotic organic compounds (XOCs) and heavy metals. XOCs and the heavy metals are hazardous and can be toxic, corrosive, flammable, reactive, and carcinogenic, among other hazards, and can also be bioaccumulative and persistent.

Given the levels of ammonium present in leachate, it is also necessary to deal with it on a hazardous level, as it is the primary pollutant of groundwater from landfills (Slack et al. 2005). This is a problem in all types of landfills, not specifically those designed for hazardous waste storage and disposal. The highly reactive environment of the leachate permits a wide range of chemical and biological transformation of the waste with the XOCs and heavy metals. These transformations in the leachate can lead to the formation of toxic byproducts from relatively harmless organic substances.

In the primarily anaerobic conditions found in capped landfills or in the lower layers of actively operating landfills, there are methanogenic conditions through redox environments. The heavy metal content of leachate reduces from acid phase to methanogenic phase. Depending on the contaminant, little or no degradation may occur in the treatment conditions of the landfill, with the toxic hazardous waste percolating down into the leachate and thereby requiring treatment. These transformations and degradation of the wastes can cause those wastes disposed of in the landfill to become even more toxic than they were at the time of disposal (Slack et al. 2005).

The design of the landfill cover system is an important consideration. A landfill cover system controls the infiltration through the top of the landfill and reduces recharge of precipitation into waste or contaminated soil; it prevents direct contact with waste or contaminated soil and also prevents fugitive emissions. Landfill cover also allows the site to be returned to some beneficial use and makes the site aesthetically acceptable to nearby residents.

Construction of a landfill results in large increases in off-site flows and sediments. Therefore, runoff from non-landfilled areas is diverted off site. The off-site flow is controlled by constructing swales and stormwater recharge ponds. Runoff from exposed excavation area is either directed to siltation basins and discharged off site or directed to localized holding sumps and sampled for contamination.

Landfill gas is generated during the natural process of bacterial decomposition of organic material contained in the wastes placed in landfills. By volume, landfill gas is about 50 percent methane and 50 percent carbon dioxide and water vapor, and also contains small amounts of nitrogen, oxygen, and hydrogen, less than 1 percent nonmethane organic compounds, and trace amounts of inorganic compounds. The decomposition process in a landfill that is generating significant amounts of gas lasts about 15 to 25 years, and the volume of gas decreases steadily over this period of time. There are two compliance options under the regulations: install a landfill gas collection system and flaring, or install a landfill gas collection system and an energy recovery system. A flare system provides the opportunity to combust the landfill gas in the event that the gas is not needed. However, landfills increasingly capture and use of landfill methane as fuel for electricity generation through the development of well fields and collection systems.

There are several treatment options for landfill leachate after collection. For example, sedimentation, air stripping, adsorption, and membrane filtration are the major physical methods used for leachate pretreatment. Coagulation with flocculation and chemical precipitation are the primary chemical methods used for treating leachate. Some physiochemical processes—including nano filtration, air stripping, and ozonation—are also utilized for COD and ammonium removal, as well as toxicity reduction (Kargi et al. 2003). Anaerobic treatment methods are suitable to concentrated waste streams such as leachate, particularly when used in an SBR (Kennedy 2000). However, leachate can also have a high variability in both strength and flows. Anaerobic treatment methods also generate significant amounts of methane, a potentially valuable product. In the SBRs, slower fill times than when treating domestic sewage are required, as they result in less stress on the biological population in the tank especially with the highest organic loading rates. Leachate from traditional municipal landfills can additionally be characterized as hazardous, especially to most wastewater systems unable to treat the high toxicity of the leachate.

REFERENCES

Chambers et al. 1991. *In situ treatment of hazardous waste-contaminated soils*. Park Ridge, NJ: Noyes Data Corporation.

Chapell, F. H. 1995. Bioremediation: Nature's way to a cleaner environment. From *U.S. Department of the Interior, U.S. Geological Survey, Fact Sheet FS-054-95*. http://water.usgs.gov/wid/html/biorem.html.

DePaoli, D., J. Wilson, and C. Thomas. 1996. Conceptual design of soil venting systems. *Journal of Environmental Engineering* 122 (5).

Eckenfelder, W. W., Jr. 2000. *Industrial water pollution control*. McGraw-Hill Series in water Resources and Environmental Engineering, New York: McGraw-Hill.

Eckenfelder, W. W., Jr., and J. L. Musterman. 1995. *Activated sludge treatment of industrial wastewater*. Lancaster, PA: Technomic Publishing Company.

Jensen, R. 1994. Successful treatment with supercritical water oxidation. *Environmental Protection* 5 (6).

Kargi, Fikret, and M. Yunus Pamukoglu. 2003. Aerobic biological treatment of pre-treated landfill leachate by fed-batch operation. *Enzyme and Microbial Technology* 333.

Kennedy, K. J., and E. M. Lentz. 2000. Treatment of landfill leachate using sequencing batch and continuous flow upflow anaerobic sludge blanket (UASB) reactors. *Water Research* 34 (14).

LaGrega, M. D., P. L. Buckingham, and J. C. Evans. 1994, 2001. *Hazardous waste management*. New York: McGraw-Hill,

Luck, F. 1999. "Wet air oxidation: Past, present and future". *Catalysis Today* 53.

Matatov-Meytal, Y.I., and M. Sheintuch. 1998. Catalytic abatement of water pollutants. *Ind. Eng. Chem. Res*. 37.

McBean, E. A., F. A. Rovers, and G. J. Farquhar. 1995. *Solid waste landfill engineering and design*. Englewood Cliffs, NJ: Prentice Hall PTR.

Metcalf and Eddy et al. 2002. *Wastewater treatment and reuse*. New York: McGraw-Hill.

Pintar, A. 2003. Catalytic processes for the purification of drinking water and industrial effluents. *Catalysis Today* 77.

Riser-Roberts, E. 1998. *Remediation of petroleum contaminated soils: biological, chemical, and physical processes*. Boca Raton, FL: CRC Press LLC.

Slack, R. J., J. R. Gronow, and N. Voulvoulis. 2005. Household hazardous waste in municipal landfills: Contaminants in leachate. *Science of the Total Environment* 337.

Stamnes, R., and J. Blanchard. 1997. *Engineering forum issue paper: Soil vapor extraction implementation experiences*. Environmental Protection Agency. Retrieved on June 1, 2008 from http://www.epa.gov/tio/tsp/download/sveissue.pdf.

U.S. EPA. 1987. *Compendium of technologies used in the treatment of hazardous waste*. EPA/625/8-87/014. Cincinnati, OH: Center for Environmental Research Information.

U.S. EPA. 1994. *Chapter II: Soil vapor extraction*. Retrieved on April 2, 2008, from http://www.epa.gov/OUST/pubs/tum_ch2.pdf.

U.S. EPA. 1999. *Wastewater technology fact sheet: Sequencing batch reactors*, May 18, 2008, from http://epa.gov/OWM/mtb/sbr_new.pdf.

U.S. EPA. 2006. *In situ treatment technologies for contaminated soil*. Retrieved on May 12, 2008, from http://www.epa.gov/tio/download/remed/542f06013.pdf.

Visvanthan, C. 1996. Hazardous waste disposal. *Resources, Conservation, and Recycling* 16.

Watts, J. R. 1997. *Hazardous Wastes: Sources, pathways, receptors*. New York: Wiley.

Weber, W. J., Jr. 1972. *Physicochemical processes for water quality control*. New York: Wiley.

Wong, J., C. Lim, and G. Nolen. 1997. *Design of remediation systems*. Boca Raton, FL: CRC Press, Inc.

Woodside, G. 1999. *Hazardous materials and hazardous waste management*, 2nd ed. New York: Wiley.

CHAPTER 7

SANITARY LANDFILL OPERATIONS

Berrin Tansel, Ph.D., P.E., F. ASCE
Florida International University
Miami, Florida

1 WHAT IS A SANITARY LANDFILL?

An economical and environmentally sustainable management of solid waste requires an integrated approach that incorporates collection, recycling, resource recovery, land disposal, and public education. Sanitary landfills are engineered systems developed for disposal of solid waste. A sanitary landfill is designed and developed in a manner that prevents pollution or harm to environment. After the landfill is completed, the land may be used for another purpose that benefits the community. Landfills are critical for most waste management strategies, because they are the simplest, cheapest, and most cost-effective method of disposing of waste (Allen 2001). However, engineering considerations should be incorporated during design, operation, closure, and postclosure of landfills so that impacts to the environment (i.e., leachate and gas releases to the environment) are minimized or mitigated. Cap, cover, liner system, and gas collection system are the basic components of a sanitary landfill. The landfill cap reduces the infiltration of precipitation while controlling leachate and gas migration. Landfill gas produced during the anaerobic biodegradation of the organic materials in the waste consists mainly of methane and carbon dioxide, with trace levels of volatile organic compounds. Pressure, concentration, and temperature gradients that develop within

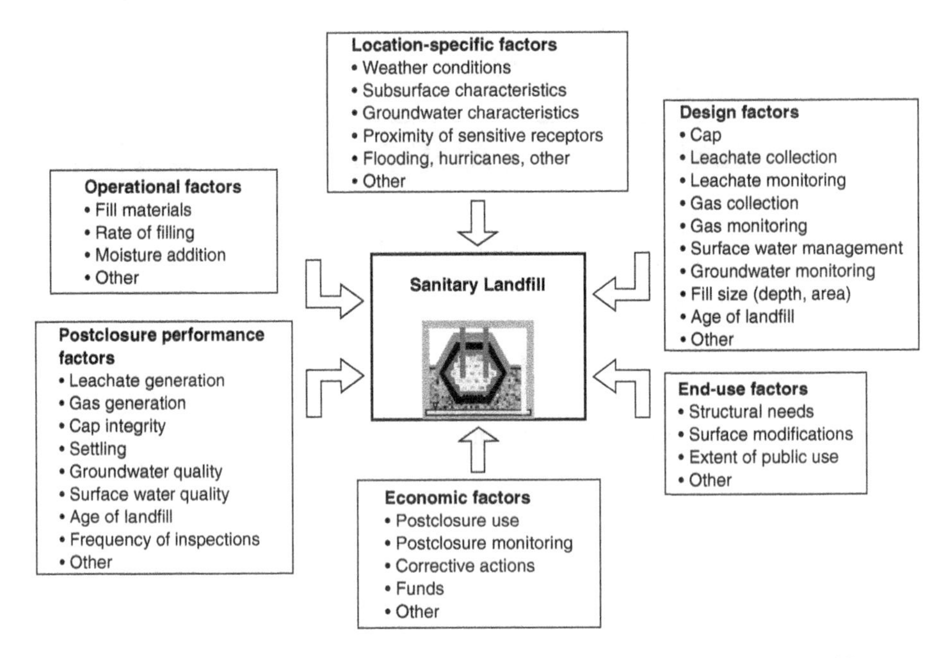

Figure 7.1 Engineering considerations for development of sanitary landfills

the landfill result in gas emissions to the atmosphere, as well as lateral migration through the surrounding soils (Nastev et al. 2001). Figure 7.1 presents the engineering considerations for development of sanitary landfills.

Design goals for sanitary landfills typically include the following (O'Leary and Tansel 1986a):

- To serve the solid waste disposal needs of a specific community or region
- To protect groundwater quality by eliminating leachate discharge
- To protect air quality and generate energy by installing a landfill gas recovery system
- To use landfill space efficiently and extend site life as much as is practical
- To minimize dumping time for site users to reduce potential nuisance conditions for neighbors
- To provide a plan for using the land after the site is closed

The general trends indicate that the number of landfills in the United States has steadily declined over the years. However, the average landfill size has increased, as shown in Figure 7.2 (U.S. EPA 2007). Since 1990, the total volume of municipal solid waste (MSW) disposed of in landfills has decreased by 4 million tons, from 142.3 million to 138.2 million tons in 2006, as presented in Table 7.1. Based on the data reported by the U.S. EPA, the net per capita discard rate (after recycling, composting, and combustion for energy recovery) was 2.53 pounds per person per day in 1960, similar to the 2.55 per capita rate in 2004, as shown in Table 7.2.

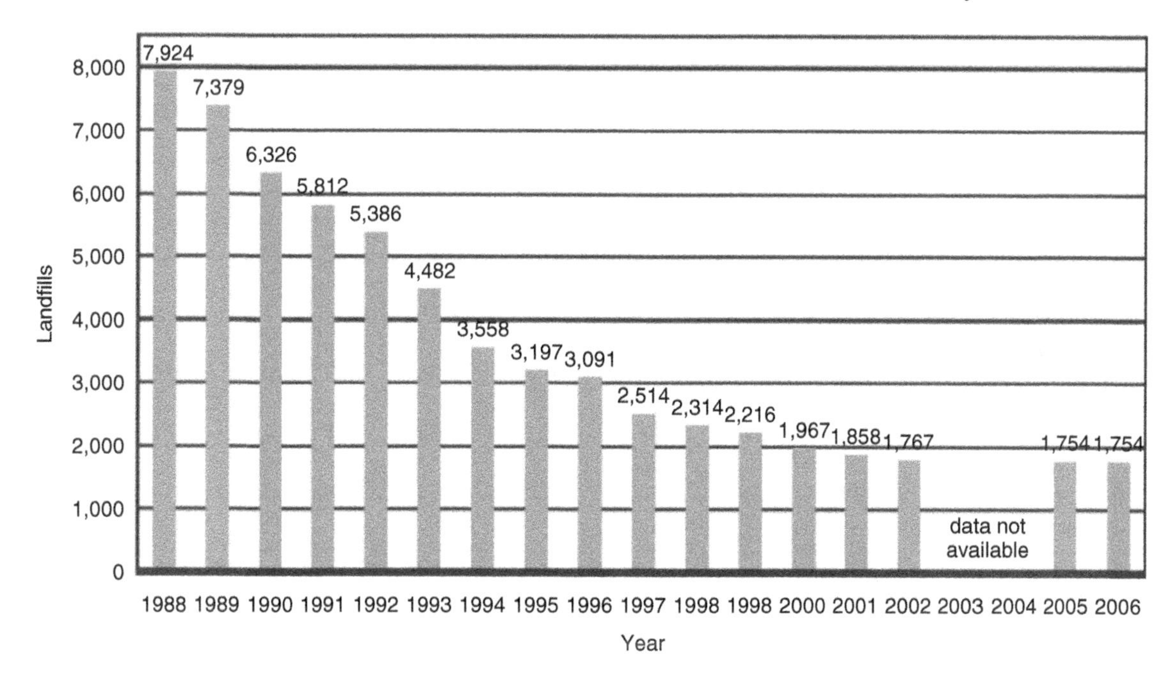

Figure 7.2 Number of landfills in the United States, 1988–2006 (U.S. EPA, 2007)

Table 7.1 Generation, Materials Recovery, Composting, Combustion with Energy Recovery, and Discards of MSW, 1960–2006 in the United States (in millions of tons)

Activity	1960	1970	1980	1990	2000	2002	2004	2005	2006
Generation	88.1	121.1	151.6	205.2	238.3	239.4	249.2	248.2	251.3
Recovery for Recycling	5.6	8.0	14.5	29.0	52.8	53.8	57.5	58.6	61.0
Recovery for composting*	Negligible	Negligible	Negligible	4.2	16.5	16.7	20.5	20.6	20.8
Total materials recovery	5.6	8.0	14.5	33.2	69.3	70.6	77.9	79.1	81.8
Combustion with energy recovery†	0.0	0.4	2.7	29.7	33.7	33.4	34.4	33.4	31.4
Discards to landfill, other disposal‡	82.5	112.7	134.4	142.3	135.3	135.5	136.9	135.6	138.2

*Composting of yard trimmings, food scraps, and other MSW organic material. Does not include backyard composting.
†Includes combustion of MSW in mass burn or refuse derived fuel form, and combustion with energy recovery of source-separated materials in MSW (e.g., wood pallets and tire-derived fuel).
‡Discards after recovery minus combustion with energy recovery. Discards include combustion without energy recovery. Details may not add to totals due to rounding.
Source: U.S. EPA (2007).

Table 7.2 Generation, Materials Recovery, Composting, Combustion with Energy Recovery, and Discards of MSW, 1960–2006 in United States (in pounds per person per day)

Activity	1960	1970	1980	1990	2000	2002	2004	2005	2006
Generation	2.68	3.25	3.66	4.50	4.64	4.55	4.65	4.59	4.60
Recovery for Recycling	0.17	0.22	0.35	0.64	1.03	1.02	1.07	1.08	1.12
Recovery for composting*	Negligible	Negligible	Negligible	0.09	0.32	0.32	0.38	0.38	0.38
Total materials recovery	0.17	0.22	0.35	0.73	1.35	1.34	1.45	1.46	1.50
Combustion with energy recovery†	0.0	0.1	0.07	0.63	0.66	0.63	0.64	0.62	0.57
Discards to landfill, other disposal‡	2.51	3.02	3.24	3.12	2.63	2.58	2.55	2.51	2.53
Population (millions)	179.979	203.984	227.255	249.907	281.442	287.985	293.660	296.410	299.398

*Composting of yard trimmings, food scraps, and other MSW organic material. Does not include backyard composting.
†Includes combustion of MSW in mass burn or refuse derived fuel form, and combustion with energy recovery of source-separated materials in MSW (e.g., wood pallets and tire-derived fuel).
‡Discards after recovery minus combustion with energy recovery. Discards include combustion without energy recovery. Details may not add to totals due to rounding.
Source: U.S. EPA (2007).

2 WASTE QUANTITIES AND TRENDS

Waste generation patterns and characteristics play an important role in planning, design, operation, and postclosure care of sanitary landfills. Waste quantities and waste composition at the source, although not controllable, can be influenced by public education, local policies, and ordinances. The hierarchy of waste management usually is as follows:

1. Waste minimization/reduction at the source
2. Recycling
3. Waste processing (with recovery of materials and energy)
4. Landfilling

Solid waste management involves the following steps:

1. *Collection*. The waste collection involves gathering and transporting solid waste to a materials processing facility, a transfer station, or a landfill.
2. *Sorting and processing of solid waste*. The sorting, processing, and transformation of solid waste materials may involve separation of bulky items, separation based on size, and separation of recyclable materials such as metals, plastic, and glass.
3. *Transfer and transport*. The transfer and transport involves the transfer of wastes to larger transport vehicle to move the waste to a processing

or disposal site if the site is located at a significant distance. The transfer of waste usually takes place at a transfer station. Transfer stations also house sorting and processing equipment to reduce waste quantities to be landfilled.

4. *Disposal*. Disposal involves the placement of the residential wastes collected and transported directly to a landfill site, left over materials from materials recovery processes, ash from combustion of solid waste, and other materials from various solid waste-processing facilities.

Proper management of solid waste requires a good estimate of the anticipated waste composition and quantities. Waste composition depends on a number of technical, environmental, social, and regulatory factors such as climate, regional characteristics, demographic characteristics, and local legislation. Waste surveys are often conducted by communities to understand the impacts of demographic factors, establish a baseline, and assess impacts of recycling programs.

The data from the surveys are used for five functions:

1. Development of a suitable management plan
2. Identification of changes and trends in composition and quantity of waste over time for future planning
3. Gathering information for the selection of appropriate equipment and technology for waste handling
4. Estimation of the amounts and types of material suitable for processing, recovery, recycling, and landfilling
5. Projection and identification of future waste management needs

3 SOURCES OF SOLID WASTE

Depending on the land-use characteristics of the areas, solid waste can be classified as follows:

- *Domestic/residential waste*. This is solid waste generated by single and multifamily household units as a consequence of household activities (e.g., cooking, cleaning, repairs, hobbies, redecoration, empty containers, packaging, clothing, old books, writing/new paper, and old furnishings). These wastes also include bulky items such as furniture and large appliances that are discarded.
- *Municipal waste*. Municipal wastes are generated from municipal activities such as street sweeping, dead animals, and abandoned vehicles.
- *Commercial waste*. Commercial waste originate from offices, wholesale and retail stores, restaurants, hotels, markets, warehouses, and other commercial establishments.
- *Institutional waste*. Institutional wastes are generated by institutions such as schools, universities, hospitals, and research institutes.

- *Construction and demolition (C&D) wastes*. Construction and demolition wastes are the waste materials generated by the construction, repair and demolition activities. C&D waste consists of primarily inert materials such as earth, stones, concrete, bricks, lumber, roofing materials, plumbing materials, heating systems and electrical wires.
- *Bulky wastes*. Bulky wastes are large household appliances such as ovens, refrigerators, and washing machines, as well as furniture, vehicle parts, trees, and branches.

4 FACTORS AFFECTING SOLID WASTE GENERATION RATES AND COMPOSITION

Historically, increases in waste generation rates correlate with increases of the gross domestic product (GDP). The waste generation rates and composition depend on nine factors:

1. Source reduction/recycling practices
2. Geographic location
3. Season
4. Use of home food waste grinders
5. Frequency of collection
6. Legislation
7. Public attitudes
8. Per capita income
9. Size of households

According to the U.S. EPA, recycling 82 million tons of MSW resulted into energy savings equivalent to more than 10 billion gallons of gasoline (U.S. EPA 2007). Recycling 1 ton of aluminum cans conserves more than 207 million BTUs, the equivalent of 36 barrels of oil, or 1,655 gallons of gasoline.

Table 7.3 presents the composition of MSW in the United States. Organic materials constitute the largest fraction of MSW. Significant amounts of material

Table 7.3 Composition of the MSW Generated in 2006

Component	Percentage
Paper and paperboard products	34 %
Yard trimmings and food scraps	25 %
Plastics	12 %
Metals	8 %
Rubber, leather, and textiles	7 %
Wood	6 %
Glass	5 %
Other miscellaneous wastes	3 %

Source: U.S. EPA (2007).

from each category was recycled or composted in 2006. The highest recovery rates were achieved in yard trimmings, paper and paperboard products, and metal products. Recycling of organic materials alone (i.e., paper and paper products) reduced the quantity of MSW disposed of in landfills and incinerated at combustion facilities by 25 percent.

5 LANDFILL PLANNING AND DESIGN

The volume of waste to be placed in a landfill can be estimated based on two factors:

1. Per capita solid waste generation rates per year
2. Projected increase in waste generation rate based on the historical records and population growth rate

The required landfill capacity for fill and development (also called air space) is significantly greater than the waste volume due to specific site development needs. The actual capacity of the landfill depends on the volume occupied by the liner system, cover material (daily, intermediate and final cover), cap system, gas collection wells, amount of waste to be deposited, in-place density of the waste, anticipated settling that the waste will undergo due to overburden stress, and biodegradation potential of the waste. The life of a landfill extends over many years and involves several phases, as follows (O'Leary and Walsh 2002):

1. *Site selection and investigation*. Potential sites for a proposed landfill are evaluated, and a public participation program is initiated to communicate with the public and minimize potential opposition to the siting process.
2. *Design and regulatory approval*. Detailed plans and specifications are prepared, regulatory approvals and financial commitments are received, and construction is initiated.
3. *Site construction*. Support facilities are constructed and first cells for solid waste deposition are developed.
4. *Operation*. The active period of the landfill during which the landfill accepts waste. The landfill typically operates 10 to 30 years, depending on the capacity of the landfill.
5. *Site closure*. The cells are filled to capacity and are closed. At the completion of waste deposition period, the landfill is capped.
6. *Long-term care*. Regulatory standards require an owner to monitor the landfill and provide facility maintenance for 30 years after closure. The postclosure care period can be longer if the site presents a risk for the community.

5.1 Site Selection

Proper planning and site selection can prevent many future design, operational, and environmental problems during the operation and postclosure periods of a landfill. During the siting process, the specific design and operational needs are identified. During this stage, important parameters must be decided (O'Leary and Walsh 2002):

- Types of wastes accepted or rejected
- Geographic area the site will serve
- Target tip fee or cost of operation
- Maximum haul distance
- Site operating life
- Profile of potential site users
- Means for coordinating with recycling and resource recovery projects

An ideal sanitary landfill location should have the following eight characteristics (O'Leary and Tansel 1986a):

1. It complies with local zoning and land use criteria, including local road weight limits and other limitations.
2. It is easily accessible by solid waste collection and transport vehicles in all weather conditions.
3. It is suitable for safely protecting surface and groundwater quality.
4. It is suitable for controlling landfill gas migration.
5. There is access to earth cover material that can be easily handled and compacted.
6. The landfill's operation is located such that it will not affect external environmentally sensitive areas.
7. There is adequate land and internal capacity to provide a buffer zone from neighboring properties, and this can be expanded.
8. The location is feasible for haul distances to user communities.
9. The location is economically feasible to acquire, develop, and operate as a landfill.

Wetlands, unstable soils, or landslide-susceptible areas, fault areas, seismic impact zones, and areas in the 100-year floodplain or in proximity to an airport are not suitable for landfill development. Land-use plans, GIS maps, floodplain maps, and aerial photographs can be used to assess suitability of the areas. Areas with a sensitive environment and areas with endangered plant or animal habitats, virgin timber land, wildlife corridors, unique physical features, and historical and archeological sites should be avoided.

Ideally, landfill sites should be located in silt and clay soils that restrict leachate and gas movement. A landfill constructed over a permeable formation such as gravel, sand, or fractured bedrock can pose a significant threat to groundwater quality (O'Leary and Tansel 1986b, 1986c). Soil characteristics are important for landfill development for three reasons:

1. *Cover material will be used to cover the solid waste daily and when an area of the landfill is completed*. The permeability of the final cover influences the quantity of leachate generated.
2. *Migration control for leachate and gas movement away from the landfill*. Impermeable soils will retard leachate and gas movement from the site. Permeable soils provide less protection and require installation of additional controls to prevent leachate and gas migration from the site.
3. *Foundation and support structures for liners, roads, and other construction*. Soils which support liner systems should be impermeable, stable and free of rocks and gravel which may adversely affect liner integrity. Soils on the road should provide the necessary traction for the vehicles and allow drainage.

The surface characteristics and subsurface formations affect the landfill's layout and drainage characteristics.

The potential sites for suitability as a landfill can be evaluated by a ranking systems based on technical, environmental, institutional (i.e., permitting), and economic criteria to identify two to four sites for more detailed evaluation for suitability as a landfill site. For these sites, additional studies are conducted to collect data to evaluate hydrogeologic characteristics, drainage patterns, geologic formations, groundwater depth, flow directions, and natural quality and construction characteristics of site soils. Also, data about existing and planned land use, surrounding land development, available utilities, highway access, political jurisdiction, and land cost are collected and evaluated. Data from soil borings below and adjacent to the proposed site are collected to determine subsurface conditions. Bore holes can subsequently be converted to groundwater wells to monitor the groundwater table fluctuations and groundwater quality.

5.2 Design and Regulatory Approval

Planning, siting, and design of a sanitary landfill requires an assessment of existing and projected needs, identification and assessment of potential sites, and preparation of a design package as listed below.

Sanitary Landfill Design Steps (Source: Conrad et al. 1981 with additions by O'Leary, Walsh, and Tansel)

1. Determine solid waste quantities and characteristics:
   a. Existing
   b. Projected
2. Compile information for potential sites:
   a. Perform boundary and topographic surveys.
   b. Prepare base maps of existing conditions on and near sites such as property boundaries, topography and slopes, surface water, wetlands, utilities roads, structures, residences, and land use.

   c. Compile hydrogeological information and preparation of location map for soils (depth, texture, structure, bulk density, porosity, permeability, moisture, ease of excavation, stability, pH, and CATION exchange capacity), bedrock (depth, type, presence of fractures, and location of surface outcrops), groundwater (average depth, seasonal fluctuations, hydraulic gradient and direction of flow, rate of flow, quality, and uses).

   d. Compile climatological data such as precipitation, evaporation, temperature, number of freezing days, and wind direction.

   e. Identify regulations (federal, state, local) and design standards applicable for loading rates, frequency of cover, distances to residences, roads, surface water and airports, monitoring, groundwater quality standards, roads, building codes, and contents of application for permit.

3. Design of filling area

   a. Select landfilling method based on site topography, site soils, site bedrock, and site groundwater.

   b. Specify design dimensions for cell width, depth, length, fill depth, liner thickness, interim cover soil thickness, and final soil cover thickness.

   c. Specify operational features such as use of cover soil, method of cover application, need for imported soil, equipment requirements, and personnel requirements.

4. Design features:

   a. Leachate controls

   b. Gas controls

   c. Surface water controls

   d. Access roads

   e. Special working areas

   f. Special waste handling

   g. Structures

   h. Utilities

   i. Recycling drop-off

   j. Fencing

   k. Lighting

   l. Wash racks

   m. Monitoring wells

   n. Landscaping

5. Prepare design package:

   a. Develop preliminary site plan of fill areas.

   b. Develop landfill contour plans. Details should include excavation plans (including benches), sequential fill plans, completed fill plans, fire, litter, vector, odor, and noise controls.

c. Compute solid waste storage volume, soil requirement volumes, and site life.

d. Develop final site plan showing normal fill areas, special working areas, leachate controls, gas controls, surface water controls, access roads, structures, utilities, fencing, lighting, wash racks, monitoring wells, and landscaping.

e. Prepare elevation plans with cross-sections of excavated fill, completed fill, and phase development of fill at interim points.

f. Prepare construction details including leachate controls, gas controls, surface water controls, access roads, structures, and monitoring wells.

g. Prepare ultimate land-use plan.

h. Prepare cost estimate.

i. Prepare design report.

j. Prepare environmental impact assessment.

k. Submit application and obtaining required permits.

l. Prepare operator's manual.

A typical cross-section of a sanitary landfill cell is presented in Figure 7.3. Design of a landfill should include specifics of the planning and development of each landfill cell. The site development and design report for sanitary landfills includes the following (O'Leary et al. 1986b):

- Site description, which includes existing site size, topography, slopes, surface water, utilities, roads, structures, land use, soil, groundwater, exploration data, bedrock and climate information

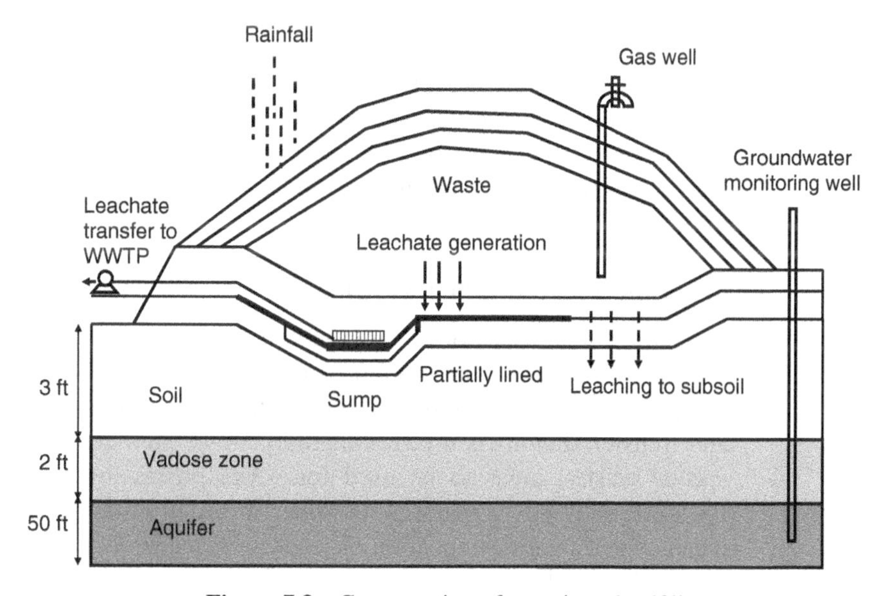

Figure 7.3 Cross-section of a sanitary landfill

- Design criteria, which include solid waste types, volumes, fill-area dimensions, and all calculations
- Operational procedures, which include site preparations, solid waste unloading, handling and covering, as well as equipment and personnel requirements
- Environmental safeguards, which include the control of leachate, surface water, gas, blowing paper, odor, and vectors

The report should also address the activities that are regulated by the U.S. EPA with regard to the following:

- Groundwater quality protection
- Landfill gas controls
- Air pollution control
- Basic operating procedures
- Safety issues
- Flood plains
- Seismic and slope stability
- Disturbance of endangered species
- Surface water discharges
- Site closure and long-term care
- Closure and long-term care financial assistance

In addition, state and local governments may require additional provisions for landfill activities relative to zoning, transportation routes (heavy loads and traffic), water discharge/water quality control, mining regulations (excavations), building permits, fugitive dust and emissions controls, and closure permits. The final use of the closed landfill should be compatible with nearby land use and the limitations of the landfill to support structures. Most closed landfills are used for recreational purposes, such as golf courses, nature preserves, or ski hills. Consideration also must be given to existing landform compatibility, settlement allowances and drainage patterns, and future residential or commercial growth in the area.

5.3 Site Construction and Landfill Layout

A landfill site consists of the area in which the waste will be filled and the areas for support facilities. The support facilities located at the site include access roads, equipment shelters, weighing scales, office space, location of waste inspection and transfer station (if used), temporary waste storage and/or disposal area for special wastes, areas to be used for waste processing (e.g., shredding), areas for stockpiling cover material and liner material, drainage facilities, landfill gas management facilities, location of leachate management system (i.e., sump), and monitoring wells.

5.3.1 Leachate Collection System

Leachate is generated due to infiltration of water into the landfill and water generated during decomposition of the waste materials. Leachate quality depends on waste composition, age of fill and stage of decomposition, temperature, moisture, and available oxygen. The concentration of COD in the leachate fluctuates significantly during the first three years after wastes are deposited and then decreases to relatively low values. Leachate control within a landfill involves prevention of migration of leachate from the fill area and landfill base to the subsoil, as well as collection and removal of the leachate from the landfill. Liner systems comprise a combination of leachate drainage and collection layer(s) and barrier layer(s). A liner system may consist of a combination of low-permeability materials such as natural clays, amended soils, and flexible geomembranes. Figure 7.4 presents the components of leachate collection systems.

Types of liner systems used at sanitary landfills include the following:

- *Single liner system*. This is a single primary barrier under the leachate collection system with an appropriate separation/protection layer.
- *Single composite liner system*. A composite liner is composed of two barriers, made of different materials, placed in intimate contact with each other to provide a beneficial combined effect of both the barriers. Usually, a flexible geomembrane is placed over a clay or amended soil barrier.
- *Double liner system*. A double liner system consists of two single liners placed one over the top of the other. This type of system offers double safety and is often used beneath industrial waste landfills. It allows the monitoring of any seepage that may escape the primary barrier layer.

The leachate collection system typically consists of a drainage layer, a perforated pipe collector system, sump collection area, and a removal system. The disposal and treatment options for the leachate collected include following:

1. Discharge to sewer lines
2. Recirculation

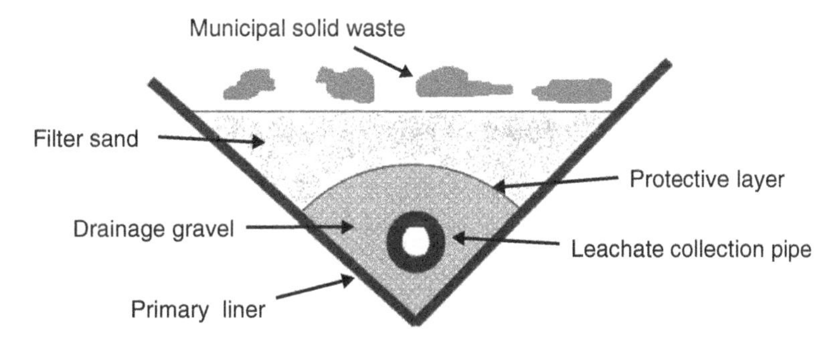

Figure 7.4 Components of leachate collection systems (adopted from Vesilind et al. 2002)

3. Evaporation of leachate
4. On-site or off-site treatment

5.3.2 Gas Collection System

Landfill gas is generated as the wastes decompose. Biological degradation of the waste initially may be aerobic due to presence of oxygen. Later, as the oxygen is depleted, the process becomes anaerobic. Methane and carbon dioxide are the principal gases produced. Figure 7.5 presents this change in gas composition in a landfill over time.

The following parameters affect the rate of decomposition and gas generation rate:

- Moisture content
- Availability of nutrients
- Absence of oxygen and toxics

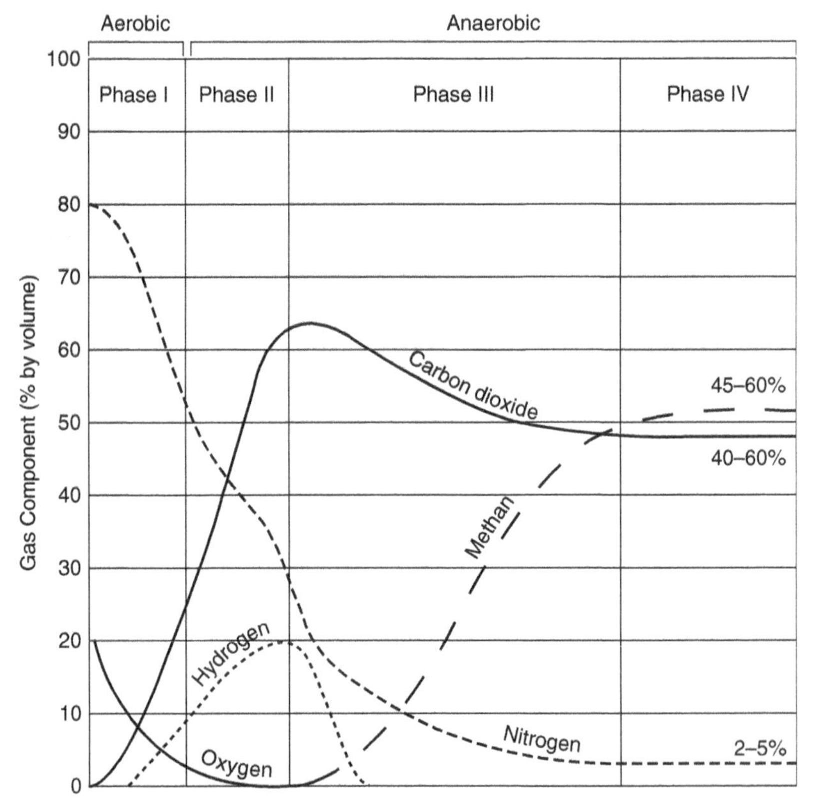

Figure 7.5 Change in gas composition in a landfill over time (U.S. EPA 1997)

- pH (Relatively neutral conditions, pH of 6.7 to 7.2, promote decomposition)
- Alkalinity and volatile acids (alkalinity greater than 2,000 mg/l as calcium carbonate and volatile acids less than 3,000 mg/L as acetic acid promote decomposition)
- Temperature (temperature between 86°F and 131°F promote decomposition)

Table 7.4 presents the typical composition of landfill gas (Vesilind et al. 2002). Landfill gases can move upward or downward in a landfill, depending on their density. For unvented landfills, the extent of this lateral movement varies with the characteristics of the cover material and the surrounding soil.

The gas management strategies may include the following options:

- Controlled passive venting
- Uncontrolled release
- Controlled collection, purification, and use

5.3.3 Surface Water Drainage System

Surface water management is needed to ensure that rainwater runoff does not drain into the waste from surrounding areas.

These objectives should be achieved by the following steps:

1. Intercept rainwater running off slopes above and outside the landfill area.
2. Collect and/or manage the rain falling on active tipping areas and channel to leachate collection drain and leachate collection sumps.
3. Divert the rainfall onto areas that have been completed and direct it to a settling pond to remove suspended silt, prior to discharge.

5.4 Operation

A landfill is developed in phases. A landfill phase is a subsection of a landfill; it consists of daily cells, lifts, daily cover, intermediate cover, liner and leachate collection facility, gas control facility, and final cover over the subarea. Landfill cells are developed from the base to the final/intermediate cover and capped.

Table 7.4 Composition of Landfill Gas

Constituent	Percentage by Volume
Methane	45–60
Carbon dioxide	40–60
Nitrogen	2–5
Oxygen	0.1–1.0
Ammonia	0.1–1.0
Hydrogen	0–0.2

Typically, each phase accommodates two to five years of solid waste volume. The phase development allows the progressive use of the landfill area, while leaving some parts undisturbed. The support facilities located at the site include access roads, equipment shelters, weighing scales, office space, location of waste inspection and transfer station (if used), temporary waste storage and/or disposal area for special wastes, areas to be used for waste processing (e.g., shredding), areas for stockpiling cover material and liner material, drainage facilities, landfill gas management facilities, location of leachate management system (i.e., sump), and monitoring wells.

The term *daily cell* is used to describe the volume of material placed in a landfill during one day. The *working face* is the area where new refuse being deposited and compacted. The working face should remain as small as possible to avoid attracting birds and creating visual problems for passersby, and to contain blowing paper. Keeping freshly deposited refuse in a well-defined and small working face is a good indication of a well-operated landfill. The minimum width of the working face or daily cell should be wide enough to accommodate the trucks or vehicles that are expected to be at the landfill at any given time. Once the working face has been completed and daily cover material provided, it is a completed cell, or *daily cell*.

The daily cell is covered with the daily cover material, usually consists of 15 to 30 cm of soil that is applied to the working faces of the landfill. The purposes of daily cover are to control the blowing of waste materials, to prevent rate, flies and other disease vectors from entering or exiting the landfill, and to control the entry of water into the landfill during operation.

A lift is a complete layer of daily cells over the active area of the landfill. Each landfill cell (or phase) consists of a series of lifts. Intermediate cover is placed at the completion of each lift. Intermediate cover is typically 45 cm. The final lift includes the cover layer. The final cover layer is applied to the entire landfill surface of the phase after all landfilling operations are complete. The final cover usually consists of multiple layers designed to enhance surface drainage, intercept percolating water, and support surface vegetation.

The following equipment is required at a landfill site (O'Leary and Walsh 2002):

- Dozers for spreading waste and daily cover
- Landfill compactors for compaction of waste
- Loader backhoes for loading waste and excavating trenches for embankment construction
- Backhoes and front-end loaders for excavating trenches and moving cover materials
- Tractor trailers for internal movement of waste or daily cover soil
- Heavy-duty backhoes for large excavation and embankment construction

5.4.1 Mitigation Measures

Mitigation measures are needed to ensure that the landfill operation does not adversely affect local environment within and outside the landfill. The mitigation measures to be implemented may include the following areas:

- *Traffic control*. Heavy traffic can interfere with local traffic. Traffic mitigation should be considered to minimize the congestion and interference with local traffic.
- *Noise control*. Adverse impacts on the local community from noise may arise. Peripheral noise abatement measures such as use of noise barriers should be considered.
- *Odor control*. Odors originating from the site could lead to complaints from the local community. Good practice options to minimize complaints from the residents include adequate compaction, speedy disposal and burial of malodorous wastes, use of daily cover, progressive capping and restoration, effective landfill gas and leachate management practices, and odor control measures while considering the prevailing wind direction.
- *Litter control*. Measures for controlling litter include consideration of prevailing wind direction, use of mobile screens close to the tipping area, and use of catch fences and netting to trap windblown litter.
- *Bird control*. Measures to mitigate bird nuisance include working in small active areas and prompt covering of waste.
- *Vermin and other pest control*. Sites with extensive nonoperational land can become infested with rabbits. The use of insecticides on exposed faces of the tipping area may control the pest problem.
- *Dust*. Dust suppression can be achieved by limiting vehicle speed and by spraying roads with water.
- *Mud control and road access management*. The site personnel need to ensure that vehicles do not carry mud off site.

All sites should have an emergency tipping area set aside from the immediate working area where incoming loads of material known to be on fire or suspected of being so can be deposited, inspected, and dealt with.

5.4.2 Monitoring

The objectives of an environmental monitoring program are to document whether a landfill is performing as designed and to ensure that the landfill is conforming to the regulatory environmental standards.

Six parameters must be monitored regularly:

1. Leachate head
2. Leachate and gas quality

3. Long-term movements of the landfill cover
4. Quality of groundwater at and around the fill areas
5. Air quality above and at the perimeter of the landfill
6. Structural integrity of the landfill cap and liner system

5.5 Site Closure and End Use

A cover or cap is an umbrella over the landfill to keep water out (to prevent leachate formation). It will generally consist of several sloped layers: clay or membrane liner (to prevent rain from intruding), overlain by a very permeable layer of sandy or gravelly soil (to promote rain runoff), overlain by topsoil in which vegetation can root (to stabilize the underlying layers of the cover). If the cover (cap) is not maintained, rain will enter the landfill, resulting in buildup of leachate to the point where the bathtub overflows its sides and wastes enter the environment.

Methodology for closing landfills can include performance-based factors as well as end-use considerations for potential threats to human health and the environment. The performance-based decision-making factors for ending PCC at landfills include (Morris et al. 2003):

- Quantification of landfill source characteristics (i.e., leachate and landfill gas)
- Definition of trends in concentrations and quantities at the source
- Evaluation and prediction of the release of constituents for potential impacts to human health and the environment
- Monitoring to confirm evaluations or predictions

In a landfill, when the moisture content of the waste is reduced to about 20 percent, rate of gas production significantly slows. With a good-quality cap design, it is possible to limit the moisture supply to a landfill, but over time, the cap may lose its integrity due to environmental and geotechnical stresses. Consequently, the cap requires periodic maintenance to prevent excessive amounts of moisture from entering the waste. The implication that monitoring will be discontinued after 30 years because the landfill is stable and no longer represents a threat to the environment requires a scientific and systematic approach for monitoring performance of closed landfills (Barlaz 2004). By implementing proper engineering measures during the operation of a landfill, the stabilization period of the MSW can be reduced significantly.

There are two types of end uses for landfills (Tansel 1998):

1. Passive use (i.e., such as green space, wildlife or nature conservancy, and hiking trails)
2. Active use (i.e., such as sports fields, golf courses, industrial uses, and transfer stations)

Table 7.5 Land Use Examples of Closed Landfills (Tansel 1998)

Land Uses
Racetrack
Baseball facility
Model airplane field
Business park and golf course
Passive parkland and a small golf course
Soccer fields, tennis courts, boat launch, fishing area, amphitheater, sledding area
Public works storage facility and transfer station
Saltwater sailing lake, golf course, wetlands, levees, amphitheater, and wildlife refuge
Snow tube park and putt-putt golf
Recreation park, wildlife refuge, and butterfly garden
Ski slopes

End use must not interfere with monitoring and other postclosure requirements. End-use features must be unaffected by subsidence, landfill gases, leachate, and erosion. Typical end uses for closed landfills are presented in Table 7.5.

5.6 Long-Term Care

After the wastes are placed in a landfill, the weight of the wastes and the soil cover cause further compression to take place. There is additional settling of completed landfill as a result of decomposition reactions. The deposited materials go through physical, chemical, and biological processes as wastes are decomposed. The physical changes are compression and settling, dissolution and transport, and absorption and adsorption. The water produced from chemical and biological reactions forms a medium for the soluble substances to dissolve and causes the unreacted materials to move.

Subtitle D of the Resource Conservation and Recovery Act (RCRA) requires a postclosure period of 30 years for nonhazardous wastes in landfills. Postclosure care (PCC) activities under Subtitle D include leachate collection and treatment, groundwater monitoring, inspection and maintenance of the final cover, and monitoring to ensure that landfill gas does not migrate off site or into on-site buildings. According to solid waste facility regulations codified in 40 CFR §258.61(b), the 30-year PCC period specified by Subtitle D can be extended or shortened by the governing regulatory agency on a site-specific basis. However, the decision to extend or shorten the postclosure care period should be based on whether the landfill is a threat to human health or the environment.

A landfill is considered functionally stable when it no longer presents an unacceptable threat to human health and the environment. The landfill activity depends on a number of factors, which include variables that relate to operations both before and after the closure of a landfill cell. Therefore, PCC decisions should be based on location-specific factors, operational factors, design factors, postclosure performance, end use, and economic considerations.

REFERENCES

Allen, A. 2001. Containment landfills: The myth of sustainability. *Journal of Engineering Geology* 60: 3–19.

Barlaz, M. 2004. *Long-term landfill management and post-closure care*. Third Intercontinental Landfill Research Symposium, Performance-Based System for Post-Closure Care at MSW Landfills—A New Approach to the Current 30-Year Time-Based System of Subtitle D, Sapporo, Japan, November 29–December 2, 2004.

Conrad, E.., J. J. Walsh, J. Atcheson, and R. B. Gardner. 1981. *Solid waste landfill design and operation practices*, EPA Draft Report, Contract No. 68-01-3915.

Morris, J., N. Durant, M. Houlihan, R. Gibbons, M. Barlaz, S. Clarke, R. Bonaparte, J. Baker, J. Gallinatti, and E. Repa. 2003. *Performance-based system for post-closure care at MSW landfills—A new approach to the current 30-year time-based system of Subtitle D, proc. of waste tech*. Environmental Research Education Foundation.

Nastev, M., R. Therrien, R. Lefebvre, and P. J. Gélinas. 2001. Landfill gas generation and migration in landfills and geological materials. *Journal of Contaminant Hydrology* 52 (1–4): 187–211.

O'Leary, P., and B. Tansel. 1986a. Sanitary landfill design: An introduction. *Waste Age* 17 (1): 120–129 (January).

O'Leary, P., and B. Tansel. 1986b. Sanitary landfill design: Landfill gas management, control and uses. *Waste Age* 17 (4): 104–118.

O'Leary, P., and B. Tansel. 1986c. Sanitary landfill design: Leachate control and treatment. *Waste Age* 17 (5): 68–85.

O'Leary, P., B. Tansel, and R. Fero. 1986a. Sanitary landfill design: How to evaluate a potential sanitary landfill site. *Waste Age* 17 (6): 78–99.

O'Leary, P., B. Tansel, and R. Fero. 1986b. Sanitary landfill design: Site preparation and plan. *Waste Age* 17 (8): 88–92.

O'Leary, P., and P. Walsh. 2002. *Landfill independent learning correspondence course materials*. Madison: University of Wisconsin, Solid and Hazardous Waste Education Center.

Vesilind, P. A., W. Worrel, and D. Reinhart. 2002. *Solid waste engineering*. Pacific Grove, CA: Brooks/Cole.

Tansel, B. 1998. Land use and development experiences with closed landfills, *The Journal of Solid Waste Management and Technology* 25 (3–4): 181–188.

U.S. EPA. 1997. Compilation of air pollution emission factors. Report Number AP-42, 5th ed. Supplement C, Office of Air Quality Planning and Statistics, U.S. Environmental Protection Agency, Washington, D.C.

U.S. EPA. 2007. *Municipal solid waste generation, recycling, and disposal in the united states: facts and figures for 2006*, EPA-530-F-07-030, Solid Waste and Emergency Response (5306P), Washington, DC 20460.

CHAPTER 8

TRANSPORTATION OF RADIOACTIVE MATERIALS

Audeen Walters Fentiman
Purdue University
West Lafayette, Indiana

1 INTRODUCTION

Transportation of radioactive materials is highly specialized and heavily regulated. Many different types of radioactive material are transported in the United States, and detailed regulations exist for transportation of each type. Radioactive materials routinely transported include isotopes used for diagnosing and treating disease, isotopes for industrial uses such as gauges to measure thickness of paper, isotopes used as tracers in research conducted in fields such as medicine, agriculture, and geology, and fuel for commercial nuclear power plants. Several types of radioactive waste are also transported. They include low-level radioactive waste from hospitals and research laboratories, transuranic waste (which includes isotopes heavier than uranium and comes primarily from laboratories responsible for making nuclear weapons), used nuclear fuel from commercial nuclear power plants, and high-level waste, which is the material remaining after used fuel from nuclear power plants is reprocessed and recycled.

This chapter will focus on the transportation of used nuclear fuel from commercial nuclear power plants, which has been called spent nuclear fuel, and high-level waste. Both of these types of waste are highly radioactive. Transporting them safely requires heavy casks that have been carefully designed and tested to protect employees at nuclear facilities, transporters, and the public from the radiation under normal and accident conditions. Since most used nuclear fuel continues to be stored at the power plants where it was used, and the United States is not currently reprocessing commercial nuclear fuel, very little used fuel or high-level waste has been transported during the first decade of the twenty-first century. However, it is possible that the United States will begin to move used nuclear fuel in the next few years. Storage capacity at the nuclear power plants is limited, and many plants are nearing their capacity. Current federal law calls for used nuclear fuel to be buried in a deep geological repository at Yucca Mountain, Nevada. Development of the Yucca Mountain repository has been delayed, and several utilities that operate nuclear plants are working together to establish an interim storage facility. In 2007 the U.S. government renewed its interest in reprocessing and may eventually build a reprocessing facility. Whether the used fuel goes to a deep geologic repository, an interim storage facility, or a reprocessing facility, it will need to be transported.

In addition, the Department of Energy has accumulated about 56,000 metric tons of high-level waste (HLW) as of April 2008, primarily at facilities in the states of Washington, Idaho, and South Carolina (USDOE 2008d). This waste is a legacy of nuclear weapons production since World War II. It is to be solidified and transported to a deep geologic repository for disposal. Casks similar to those designed for used nuclear fuel will be required for the transportation of HLW.

Because many people are unfamiliar with the composition and characteristics of radioactive waste, section 2 of this chapter will define radioactive materials, provide an overview of the nuclear fuel cycle, describe used nuclear fuel, briefly explain the reprocessing of used fuel, and describe high-level waste. An understanding of the material to be transported is important if the reader is to have an appreciation for the transportation vehicles and regulations.

Section 3 discusses the amount and location of used nuclear fuel and high-level waste awaiting disposal and locations to which these wastes will be transported under various scenarios. The Office of Civilian Radioactive Waste Management (OCRWM) within the Department of Energy is responsible for developing and managing the transportation system for used nuclear fuel and high-level waste. OCRWM's structure, its transportation plan, and transportation options will be described in section 4. Regulations governing the transportation of used nuclear fuel and high-level waste (HLW) are covered in section 5. Section 6 is devoted to a description of the casks that have been used to transport used nuclear fuel and HLW. For completeness, a brief description of the transportation methods and requirements for other types of radioactive materials will be presented in section 7. The chapter concludes with acknowledgments and references.

2 RADIOACTIVE MATERIAL, USED NUCLEAR FUEL, AND HIGH-LEVEL WASTE

2.1 Radioactive Material and Radiation

Radioactive material is material that includes some radioactive atoms. A radioactive atom is one whose nucleus is unstable and gives off energy in the form of radiation in an effort to reach a stable configuration. Several types of radiation can be emitted, but the most common ones in high-level radioactive waste are alpha particles, beta particles, gamma rays, and sometimes neutrons. These types of radiation are called *ionizing radiation* because they interact with matter by knocking electrons out of their orbits and forming ions in the matter through which they pass. Large amounts of ionizing radiation passing through a living organism can ionize enough atoms to cause damage or even death. Thus, it is important to protect people and other living things from highly radioactive material when transporting it.

People are protected from radiation by placing a substance around the radioactive material to absorb the radiation. The general term for this protective layer is *shielding*, and many substances can be used as effective shielding. Alpha particles, which consist of two neutrons and two protons, are not very penetrating and can be stopped by a sheet of paper. Beta particles, which are electrons, can be stopped by less than a sixteenth of an inch of aluminum. Most gamma rays, which are more penetrating than alphas or betas, can be stopped by 6 to 8 inches of steel or a greater thickness of concrete. Neutrons are most effectively stopped by materials made of molecules that contain hydrogen, such as water or plastic. Figure 8.1 illustrates the penetrating power of the various types of radiation. Transportation casks for highly radioactive material must be made of materials that will protect people from all types of radiation being emitted.

2.2 Nuclear Fuel Cycle and Used Nuclear Fuel

Commercial nuclear power plants in the United States are fueled with uranium, a naturally occurring radioactive metal found in the earth's crust. To prepare fuel for the reactor, the uranium is mined, processed, enriched, and fabricated into fuel elements. After the fuel has been used in the reactor, it is removed and stored on site. Eventually, it will be buried or reprocessed and recycled.

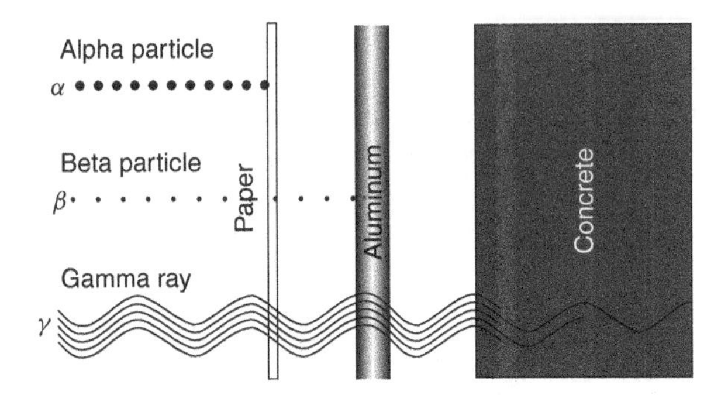

Figure 8.1 Types of radiation and shielding (Fentiman 2008)

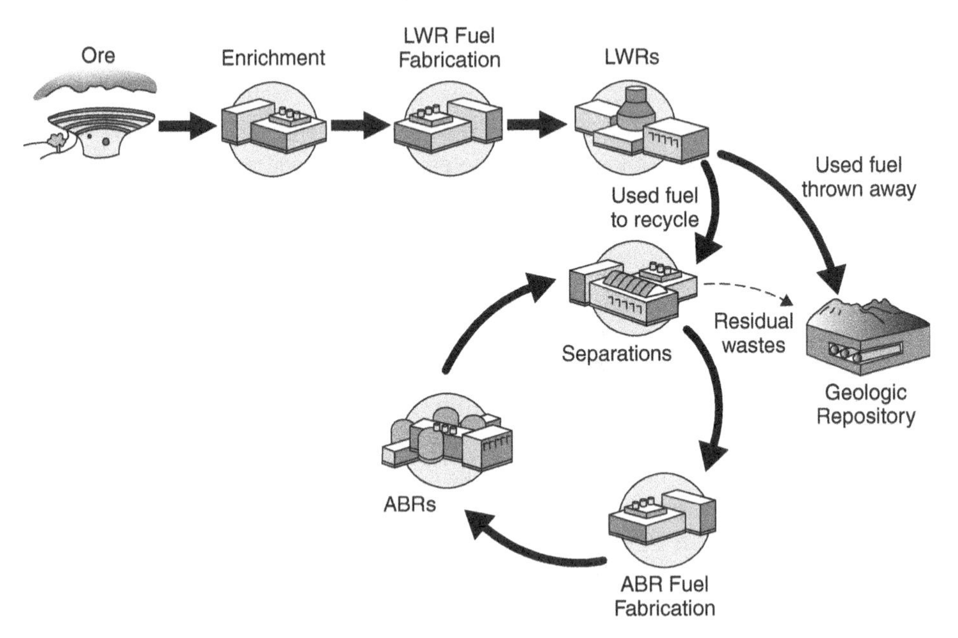

Figure 8.2 Open and closed fuel cycles (Courtesy of the Department of Energy) (USDOE Office of Nuclear Energy 2008)

If the used nuclear fuel is buried, the fuel cycle is said to be open, but if the used nuclear fuel is reprocessed and recycled, the fuel cycle is said to be closed. Figure 8.2 shows the nuclear fuel cycle. A short description of the steps in the fuel cycle will help the reader to understand the nature of the used nuclear fuel that will be transported to the repository or the reprocessing facility.

On average, uranium is found in concentrations of 2.8 parts per million in the earth's crust (World Nuclear Association 2008). However, like most metals, uranium is found in rich deposits in some locations. The countries with the largest known recoverable uranium resources are shown in Table 8.1.

Table 8.1 Known Recoverable Resources of Uranium

Country	Tonnes U	Percentage of World's Uranium
Australia	1,143,000	24%
Kazakhstan	816,000	17%
Canada	444,000	9%
United States	342,000	7%
South Africa	341,000	7%
Namibia	282,000	6%

Source: World Nuclear Association 2008.

Uranium is typically mined from the ground in one of three ways:

1. Shaft mines—which are no longer used in the United States
2. Open pit mines or strip mines
3. In-situ leaching or solution mines

Shaft mines for uranium, and other materials such as coal, have proven to be expensive and dangerous for miners. Open pit mines are more cost-effective if the ore body is close to the surface, but they leave scars on the earth. In solution mining, a solution designed to dissolve the uranium is pumped into the ore body, and when the uranium has dissolved, the solution is pumped back to the surface. This process does not disturb the surface, but some are concerned about to potential for groundwater contamination. After it is mined, the uranium is processed and the resulting material is U_3O_8, which is known as *yellow cake*.

Naturally occurring uranium consists primarily of two isotopes, U-235 (0.71 percent) and U-238 (99.29 percent). U-235 is the isotope that fissions and generates the power in a nuclear reactor. For power plants such as those in the United States to operate efficiently, the uranium fuel should be about 5 percent U-235. A process called *enrichment* is used to increase the U-235 content from 0.71 to 5 percent. The enriched uranium is converted to UO_2 and pressed into pellets, which are, in turn, loaded into fuel pins. The fuel pins are assembled into bundles and put into the reactor. Figure 8.3 shows a nuclear fuel assembly with spaces for fuel pins.

In the reactor, the U-235 fissions (splits) when it is struck by a neutron, resulting in two nuclei known as fission products, a few neutrons, and energy. The neutrons that are released go on to strike other U-235 atoms, creating a chain reaction that yields a steady supply of energy. When almost all of the U-235 atoms have fissioned, the fuel can no longer sustain a chain reaction and must be replaced. This used fuel is removed from the reactor and stored in a deep pool of water at the reactor site. Eventually, it will be transported to a central storage facility, a permanent repository, or a reprocessing/recycling facility.

Figure 8.4 shows the typical composition of used nuclear fuel. It contains about 96 percent uranium, which is a mixture of U-238 and some U-235 atoms

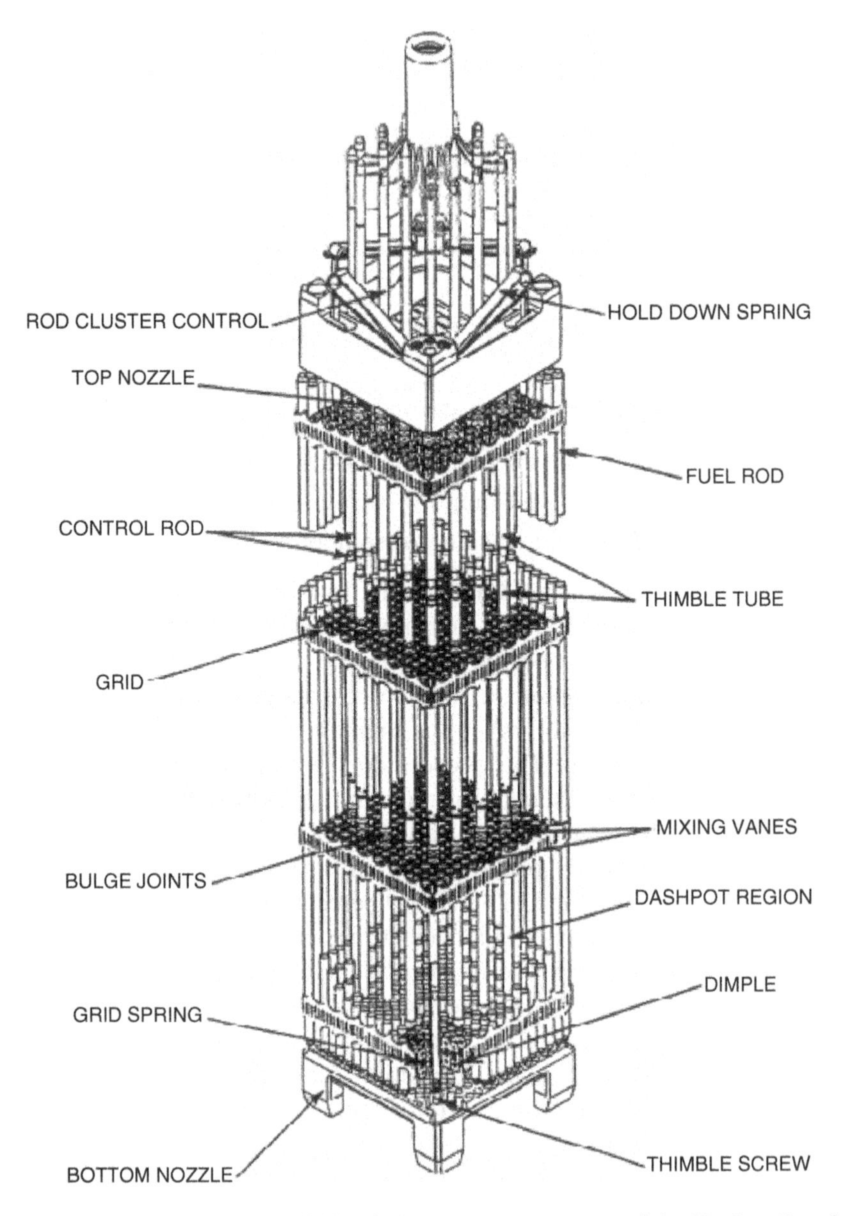

Figure 8.3 Light water reactor (PWR) fuel assembly (Courtesy of the Nuclear Regulatory Commission)

that did not fission. The uranium is radioactive but emits mostly alpha and beta particles and low-energy gamma rays. Not much shielding would be required to absorb the radiation from uranium. Approximately 3 percent of the used fuel is fission products, the two nuclei formed when U-235 fissions. Some of these are radioactive, emitting very energetic gamma rays and accounting for most of the

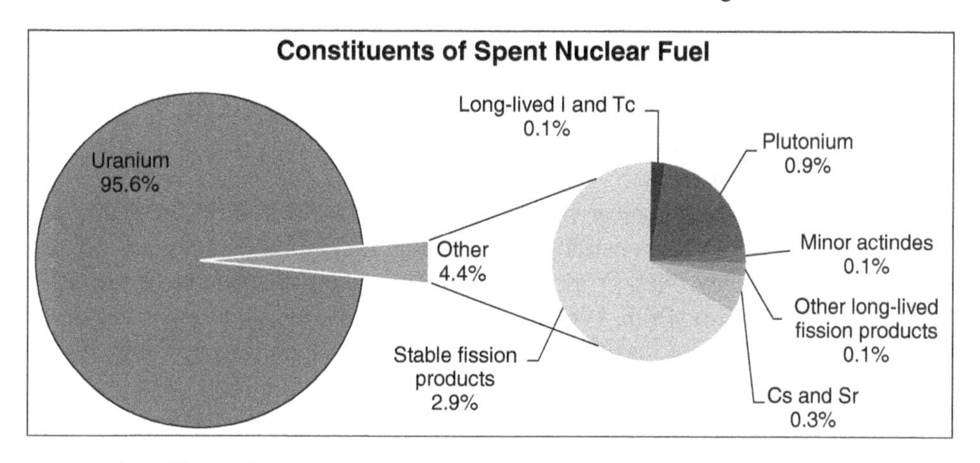

Figure 8.4 Constituents of spent nuclear fuel (Ryskamp 2003).

high levels of penetrating radiation emitted by used nuclear fuel. Because of the fission products present, casks for transporting used nuclear fuel must have thick shielding. The remaining 1 percent of the used fuel is plutonium. Plutonium is formed when a neutron released during fission is absorbed by a U-238 atom. The probability of such an absorption is low, but the absorption does occasionally occur. Like uranium, plutonium emits primarily alpha and beta particles, but it does emit some energetic gamma rays that require shielding.

2.3 Reprocessing, Recycling, and High-Level Waste

When nuclear power was introduced in the United States in the late 1950s, developers expected to reprocess and recycle the used fuel, resulting in a closed fuel cycle. There are three reasons for reprocessing and recycling:

1. The uranium in the used fuel could be utilized.
2. Plutonium, much of which fissions like U-235, could be used to generate power.
3. Less than 5 percent of the used fuel would be sent to the repository, vastly reducing the amount of repository space needed.

Reprocessing technology was developed in the United States in the 1940s, and several reprocessing methods have been used. The PUREX (Plutonium Uranium Extraction) process was used by the U.S. government to reprocess used fuel taken from reactors that were run by the federal government (not commercial reactors) with the goal of extracting plutonium and using it to make nuclear weapons. In the PUREX process, used nuclear fuel is chopped up and dissolved in concentrated nitric acid. The uranium and plutonium are removed and the remaining 3 percent of the used fuel, which is still dissolved in the nitric acid solution, is pumped to underground tanks. To prepare the high-level waste for disposal, the liquid will be evaporated, and the remaining calcine will be mixed with a glass frit, melted, and poured into canisters where it will be allowed to cool. The canisters are expected to be about the size of used nuclear fuel bundles from commercial nuclear power plants and are to be buried in the deep geologic repository.

Used nuclear fuel from commercial nuclear power plants was not reprocessed until a demonstration commercial fuel reprocessing plant was constructed in West Valley, New York, in 1966. That plant operated as a nuclear fuel reprocessing facility from 1966 to 1972 before being shut down because it was economically not feasible (USDOE 2008e). U.S. policy shifted in 1977, and reprocessing of commercial used nuclear fuel was prohibited because of worries about the security of plutonium that had been separated from the rest of the used fuel. In 2007, the policy shifted again. The U.S. government proposed the Global Nuclear Energy Partnership (GNEP), under which countries with used fuel reprocessing capabilities would reprocess fuel for countries that did not have such capabilities (GNEP 2007). One goal of GNEP was to reduce the chances that plutonium from reprocessed used fuel can be diverted. As part of the GNEP program, Argonne National Laboratory began developing a new reprocessing methodology, known as UREX+1a, which does not isolate plutonium. The process also removes some of the fission products that can be treated separately, rather than being sent to the repository, thus further reducing the required repository capacity.

The composition of high-level waste from commercial used nuclear fuel reprocessed in the future will depend on the reprocessing methodology selected. The waste is likely to contain some of the fission products and trace amounts of uranium and plutonium. HLW currently stored at nuclear weapons facilities contains many of the fission products that emit high-energy gamma rays. As a result, casks for transporting HLW will need to provide shielding similar to that required for used nuclear fuel.

3 AMOUNTS AND LOCATIONS OF USED NUCLEAR FUEL AND HLW TO BE TRANSPORTED

There are currently 104 nuclear power plants operating in the United States, generating about 20 percent of the nation's electricity. Figure 8.5 shows the location of these plants. Each plant replaces about one-third of its fuel every 18 to 24 months. After being removed from the reactor, the used fuel is first sent to a spent fuel storage pool at the nuclear power plant. According to the Nuclear Waste Policy Act of 1982, the federal government was to begin removing that fuel from storage pools and disposing of it in a deep geologic repository in 1998. Since the repository is not ready, and is not widely expected to be ready before 2020 (C&E News 2008), used nuclear fuel is accumulating at the nuclear power plants. At some sites, the spent fuel pools are full. At those sites, the oldest used fuel is being stored in dry storage casks on site. The total amount of used nuclear fuel from commercial power plants awaiting disposal was about 47,000 metric

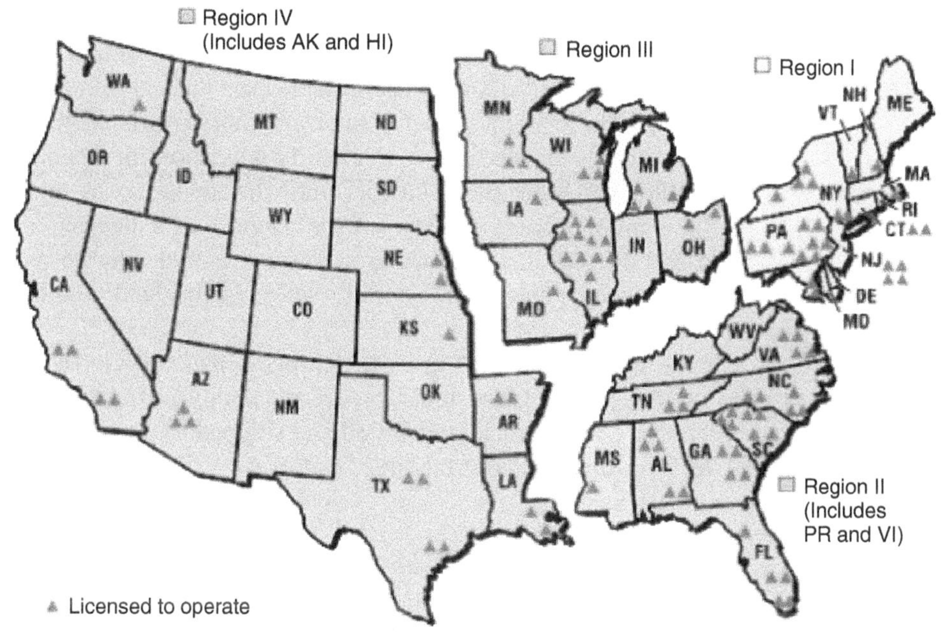

Figure 8.5 Locations of operating nuclear reactors (Courtesy of the Nuclear Regulatory Commission) (USNRC 2008a)

Table 8.2 High-Level Waste Inventory

Site	2005 Inventory (cubic meters)
Hanford	209,940
Idaho National Laboratory	8,035
Savannah River Site	132,234

Source: U.S. DOE's Central Internet Database (CID), 2008

tonnes in 2002, the latest year for which data are available (USDOE Energy Information Agency 2004).

Most of the HLW is stored in underground tanks at nuclear weapons facilities. Table 8.2 lists the locations of the largest amounts of HLW and the number of cubic meters of waste stored at each facility. The liquid waste and sludge will be dried and mixed with borosilicate glass in preparation for disposal in a geologic repository.

The size and complexity of the transportation system required to move the used nuclear fuel and HLW will depend on the fuel cycle option the United States chooses to pursue: once-through or reprocessing with recycling. If the U.S. government continues to require a once-through nuclear fuel cycle, all of the used nuclear fuel at the commercial nuclear plants and the HLW at the government-operated reprocessing facilities will be sent to a deep geologic repository for disposal. Under this scenario, the used nuclear fuel could be moved more than once. The owners of some nuclear power plants with full, or nearly full, spent fuel pools are attempting to identify a location for a central storage facility, where they can store the used fuel until the repository is opened. If a location for the central storage facility can be found and a license can be obtained to build and operate it, some of the used nuclear fuel may be moved twice—first from the power plant to the central storage facility and then from storage to the repository.

If the United States chooses to reprocess and recycle the used nuclear fuel, the used fuel will go to the reprocessing facility. The HLW generated during reprocessing would then be vitrified and shipped to the deep geologic repository. HLW from the weapons facilities and West Valley would also go to the deep geologic repository.

4 RESPONSIBILITY FOR TRANSPORTING USED NUCLEAR FUEL AND HLW

The Nuclear Waste Policy Act of 1982, amended in 1987, was an Act "to provide for the development of repositories for the disposal of high-level radioactive waste and spent nuclear fuel, to establish a program of research, development and demonstration regarding the disposal of high-level radioactive waste and spent nuclear fuel, and for other purposes" (USDOE 2004). Among those other purposes were establishing a fund to pay for the disposal of the waste, including transportation, and establishing the Office of Civilian Radioactive Waste Management (OCRWM) within the Department of Energy to oversee transportation and disposal of the waste (OCRWM 2004).

Since 1983, the Nuclear Waste Fund has collected 1.0 mill ($0.001) for each kilowatt-hour of electricity generated using nuclear fuel. The amount collected from the 1 mill per kilowatt-hour fee was $24.9 billion as of March 2005 (Bradish 2005). The Department of Defense contributes to the Nuclear Waste Fund to cover the cost of disposing of HLW from its facilities. Approximately $8.9 billion from the Nuclear Waste Fund had been spent through 2005, most of it to conduct research at and prepare the license application for a geologic repository at Yucca Mountain (Bradish 2005).

Since it was established, OCRWM has issued several documents related to plans for transportation of used nuclear fuel and HLW. The most recent document is a strategic plan released in November 2003 (OCRWM 2003). The plan notes that there have been approximately 3,000 shipments of used nuclear fuel and HLW in the United States, all of which have been made safely. In addition, there has been "extensive worldwide experience" with spent nuclear fuel transportation.

The system for transporting waste to a deep geologic repository will build on that U.S. and foreign experience.

Issues to be considered in developing the transportation system include:

- Selection of transportation routes and modes
- Emergency response planning and training
- Safeguards and security
- Operational practices
- Communications and information access
- Waste packaging for transportation
- Worker protection, training, training standards, and qualifications (OCRWM 2003)

These issues are to be discussed in detail with the states and tribes through whose land the waste will be transported.

Section 137 of the Nuclear Waste Policy Act requires the Department of Energy to use private contractors as much as possible in transporting radioactive waste to the deep geologic repository. According to the OCRWM plan, private contractors will manufacture the shipping casks, operate the transportation system, and provide maintenance services.

The transportation system is likely to require capacity to ship radioactive waste in trucks and by rail. In April 2004, the Department of Energy decided that most of the radioactive waste would be transported to the repository by rail using dedicated trains—meaning trains whose only cargo is the radioactive waste (OCRWM 2008). Reasons for selection of a rail system include increased safety and security, as well as decreased operating costs. Nuclear fuel assemblies are very heavy (uranium is denser than lead), and the casks used to transport them must be made of dense metal several inches thick to provide adequate radiation shielding. Because of the weight limits on trucks, each truck can carry only a few used nuclear fuel assemblies, typically one to nine, depending on the design of the fuel assemblies. A rail cask can carry nearly 10 times as many assemblies. Transporting the used nuclear fuel by rail will result in fewer shipments. Some truck casks will be needed because not all nuclear power plants are currently accessible by rail, and trucks will be required to transport the used fuel from the power plant to the nearest railroad. As of 2008, there was no rail access to Yucca Mountain. Current railroads can get the used nuclear fuel to within about 255 miles of the proposed repository. The Department of Energy is considering plans to construct a rail line from either Caliente or Mina, Nevada, to Yucca Mountain, but a rail line has not been approved and construction has not begun (OCRWM 2008).

One other component of the transportation system needs to be discussed. It is the Transportation, Aging and Disposal Canister (TAD). Currently, used nuclear fuel is expected to be stored at the reactor in one type of cask, transported to the repository in another, and disposed of in yet another. The cask used for each function must meet a specific set of criteria. For example, the transportation cask must be capable of withstanding several types of accidents while the disposal cask must be highly corrosion resistant. Different types of materials are used to meet those requirements, and it would be prohibitively expensive to build a cask that met the requirements for all three uses. In addition, U.S. nuclear power plants have many different designs, and those plants use fuel assemblies with different dimensions. Rather than design and build several different casks to store, transport, and dispose of these different shapes and sizes of fuel assemblies, OCRWM plans to build a canister whose internal structure can be varied to accommodate a variety of assembly designs but whose external dimensions allow it to be put into a standard cask. The TAD also makes it easier to transfer the used fuel assemblies from storage cask, to transportation cask, to disposal cask with minimum effort and risk. Five-year contracts for the design, licensing, and demonstration of the TAD canister system were awarded to two corporations in May 2008 (USDOE 2008b).

5 REGULATIONS GOVERNING TRANSPORTATION OF USED NUCLEAR FUEL AND HLW

The U.S. Nuclear Regulatory Commission (NRC) and the U.S. Department of Transportation (DOT) regulate the transportation of used nuclear fuel and HLW. The NRC's regulations, found in 10 CFR Part 71, specify the requirements for packaging for these highly radioactive materials, including procedures for approving the packaging. Packages used for transporting used nuclear fuel and HLW are referred to as Type B packages. Type A packages are used for radioactive materials emitting much less radiation than used nuclear fuel or HLW. Type B packages are heavy casks designed to shield people and the environment from radiation emitted by the used fuel or HLW and to contain the radioactive material under both normal and accident conditions.

Any packages used for transporting used nuclear fuel must have a Radioactive Material Package Certificate of Compliance from the U.S. Nuclear Regulatory Commission. To obtain a Certificate of Compliance, the cask manufacturer must first submit an application providing information specified in NUREG-1617, "Standard Review Plan for Transportation Packages for Spent Nuclear Fuel" (USNRC 2000). The application must address the safety and operational characteristics of the package, including design analysis for structural and thermal characteristics, radiation shielding, nuclear criticality, material content confinement, and the four accident test conditions listed. In addition, the application must contain operational guidance, such as any testing and maintenance requirements, operating procedures, and conditions for package use (USNRC 2007b). Any cask for transporting used nuclear fuel must contain the radioactive material under normal and accident conditions, provide both heat and radiation shielding, and prevent nuclear criticality (10 CFR 71).

The U.S. Nuclear Regulatory Commission (NRC) also regulates the physical protection of shipments of commercial used nuclear fuel. The NRC approves routes for transporting used nuclear fuel on the highways. All routes must meet the Department of Transportation regulations specified in 49 CFR 397.10. After the route is approved by the NRC, a company making a shipment of used nuclear fuel along that route must fulfill several requirements (USNRC 2007a):

- Notify the NRC of the shipment.
- Have procedures in place for dealing with any emergency during the shipment.
- Notify the governor or governor's designee of each state through which the shipment will pass.
- Work with local law enforcement agencies to make arrangements for shipping through their jurisdictions.
- Provide armed guards for the shipment in densely populated areas.

At the present time, there is no specific guidance for selecting rail routes for shipments of used nuclear fuel, but regulations for selecting rail routes for shipments of hazardous materials must be met.

Department of Transportation regulations govern selection of the truck routes for used nuclear fuel. The route selection methodology is to first identify the objective—which, in the case of transportation of used nuclear fuel is to minimize impact of the shipment, under both normal and accident conditions, on, for example, public health and safety, the environment, and property. The next steps are to develop a list of metrics (comparison factors) to consider when determining the shipment's impact, identify alternate routes, evaluate those routes in light of the metrics to be considered, and select the route that minimizes impact.

For shipments of used nuclear fuel, both primary and secondary route comparison factors are identified. In addition, both radiological and nonradiological impacts must be considered. Primary comparison factors include radiation exposure under normal and accident conditions and economic risk due to accidental release of radioactive material. Secondary comparison factors include availability of emergency response teams, ability to evacuate and area, presence of facilities that cannot be easily evacuated such as prisons or hospitals, and potential damage as a result of traffic accidents. The final secondary comparison factor, damage as a result of traffic accidents, is a nonradiological factor.

Sandia National Laboratories has developed a very detailed computer program called RADTRAN that is used to evaluate the risk associated with shipments of used nuclear fuel along potential routes. RADTRAN takes into account several factors for each route to be evaluated, including traffic pattern data and accident rate data, stopping points along the route, type of package used, composition of the material being transported, and population distribution along the route. It calculates doses to people living along the route, people traveling along the same route—both in the same direction as the shipment and in the opposite direction—and the crew transporting the used fuel.

The Department of Transportation also regulates the labeling of shipments of radioactive materials, including used nuclear fuel and HLW. There are three levels of labels used for packages containing radioactive materials, with the level of the label being determined by the maximum dose at any point on the surface of the cask. *White-I* label is used if the surface dose is less than 0.5 millirem/hour (mrem/hr). *Yellow-II* label is used if the surface dose is between 0.5 mrem/hr and 50mrem/hr, and *Yellow-III* is used if the dose is greater than 50 mrem/hr but less than 200 mrem/hr or if the material being shipped is a "highway route controlled quantity" (49 CFR 172). Used nuclear fuel and HLW are referred to as "highway route controlled quantities," and shipments must carry the Yellow-III label even though the dose from a shipping cask loaded with used nuclear fuel must be no more than 10 mrem/hr at 2 meters from the cask surface (49 CFR 172).

Finally, drivers of trucks transporting radioactive materials must typically meet several requirements. No specific requirements have yet been set for drivers of trucks loaded with used nuclear fuel. However, the Department of Energy has published requirements for drivers of trucks carrying shipments to the Waste Isolation Pilot Plant, a disposal site for transuranic waste (a type of radioactive waste) near Carlsbad, New Mexico. Those requirements include experience (minimum of 325,000 miles driven in the past five years), a stringent background check, and taking and passing over 20 training programs on topics related to safe handling and transportation of radioactive materials.

6 TRANSPORTATION CASKS FOR USED NUCLEAR FUEL AND HLW

This section describes the factors that must be considered when designing a cask for transporting used nuclear fuel or high-level waste, the process for licensing those casks, and the tests the casks must pass in order to receive a certificate of compliance. Some casks for transporting used nuclear fuel do exist, have received certificates of compliance, and have been used to transport used nuclear fuel in the United States. Information on those casks and their current status will be provided.

Figure 8.6 is an illustration of a generic truck cask for shipping used nuclear fuel. A truck cask typically weighs about 50,000 pounds and is about 4 feet in diameter and 20 feet long. Its cargo, usually one to nine used nuclear fuel assemblies, is contained in the inner basket, which holds the fuel assemblies firmly in place during transportation. The cask has inner and outer steel shells with a layer of lead between them. The steel provides structural strength, while

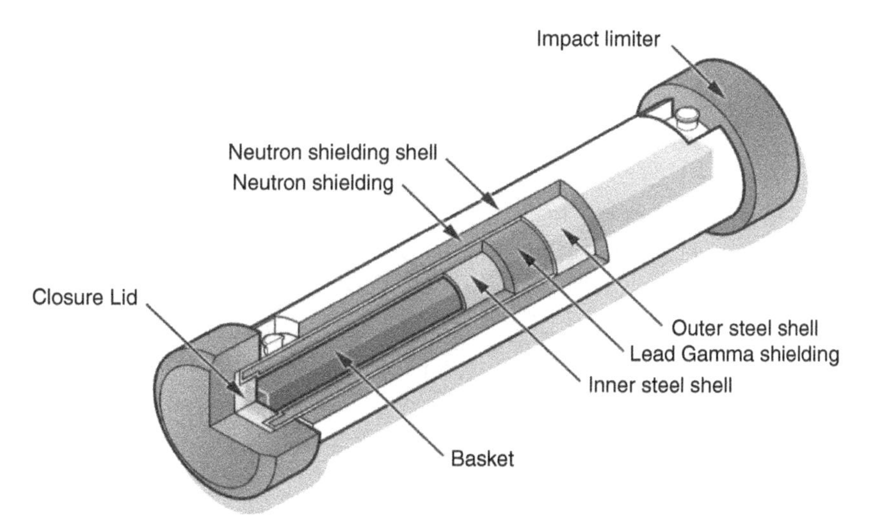

Figure 8.6 Generic truck cask (Courtesy of the Nuclear Regulatory Commission)

both the steel and the lead provide shielding that absorbs gamma rays emitted by the used fuel. The closure lid is actually a very complex component, usually composed of an inner and outer lid and many fasteners. It must be removed to load and unload the used fuel, but it must also provide an air-tight seal during transportation. The outer neutron shielding shell is typically made of a polymer that has high hydrogen content. Water is sometimes used. On the ends of the cask are impact limiters. Made of balsa wood or some synthetic material that will collapse slowly when struck, the impact limiter is designed to cushion the cask in the event of an accident.

Figure 8.7 is an illustration of a generic rail cask for shipping used nuclear fuel. It is similar in structure to the truck cask but can weigh about 250,000 pounds, have a diameter of 8 feet, and be about 25 feet long.

In 1998, the Oak Ridge National Laboratory issued, for the Department of Energy, a document titled "The Radioactive Materials Packaging Handbook" (Shappert 1998). It provided, in one document, the information that a company designing and building a Type B cask would need to manufacture a cask that could meet all applicable NRC and DOT requirements. Authors of the document took into account not only the regulations but also fabrication and operational experience with earlier Type B casks, since any difficulties with fabrication and operation can often be overcome by design modifications.

Factors to be considered in cask design include the characteristics of the material to be shipped, not only the type and amount of radiation emitted but also the chemical and physical form of the material as well as its mass, density, and volume. In addition, heat generated by the used fuel must be considered. The total number of shipments to be made in any one cask is also important. If each

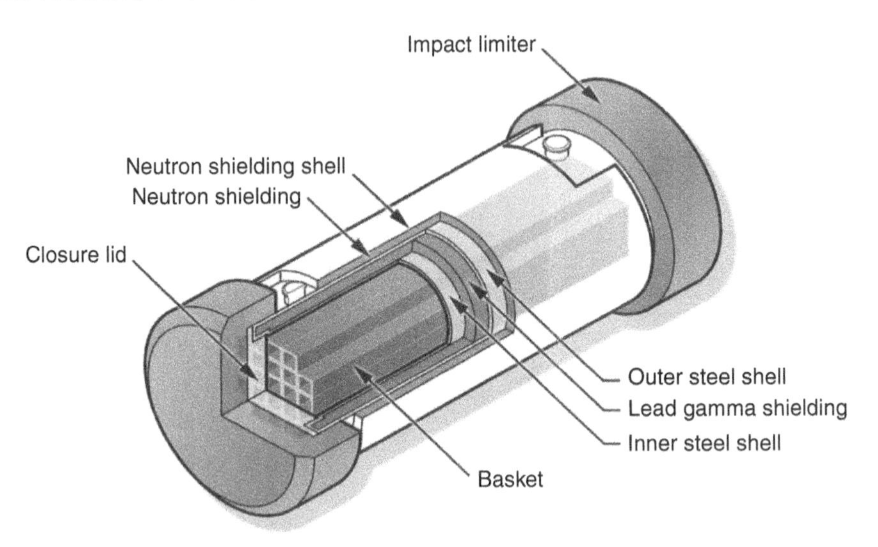

Figure 8.7 Generic rail cask (Courtesy of the Nuclear Regulatory Commission)

cask is to be used over a long period of time, normal wear and tear on the cask, operational procedures, and maintenance requirements need to be considered in the design phase. The cask must be designed to meet NRC and DOT performance requirements under normal and accident conditions. All of these factors are taken into account in selection of the composition and thickness of the cask body, lid, and impact limiters.

Another consideration in the design of the cask is the interface with existing facilities where used fuel is currently stored and the facilities to which it will be taken. The lifting trunnions, lid and fasteners, supports and tie-downs must be compatible with the handling equipment at the places where the used nuclear fuel is going to be loaded and unloaded. Since there are many different types of nuclear power plants operating in the United States and many different sizes of nuclear fuel assemblies being used, some effort will be required to design a transportation system that can be compatible with all of them. In addition, the shipping casks will have to be compatible with the handling equipment at the repository, interim storage facility, or reprocessing facility.

As mentioned earlier, over 3,000 shipments of used nuclear fuel and high-level waste have been made in the United States. Obviously, the casks to transport this material exist. Several different types of casks have been built and currently have certificates of compliance. A few of these casks and their characteristics are shown in Table 8.3. Although the current casks meet the federal requirements and serve for transporting a few used fuel assemblies, a fleet of casks with a standard design will be needed to move the used fuel once the federal government decides whether to store, bury, or reprocess/recycle it.

Table 8.3 Representative Currently Licensed Casks and Their Properties

Cask Name and Certificate of Compliance Retrieval Number	Company	Capacity	Weight Loaded	Design Specs	Rail or Truck	Comments
NAC-STC 1019235	NAC International	36 PWR fuel assemblies	125 tons loaded	Stainless steel, lead and polymer shielded	Rail	Can be used for storage or transportation
NLI-1/2:1019010	NAC International	1 PWR assembly, 2 BWR assemblies, consolidated LWR fuel, or metallic rods	49,250 pounds (22,340 kilograms) loaded	Depleted uranium, water, and lead-shielded shipping cask, encased in stainless steel, neutron shield of(borated) water-ethylene glycol mixture	Truck	
NLI-10/241019023	NAC International	10 PWR or 24 BWR fuel assemblies	194,000 pounds loaded	Lead, water, depleted uranium and high temperature polymer shield, encased in stainless steel	Rail	
GA-41019226	General Atomics	4 (or 9, respectively) PWR spent fuel assemblies	55,000 pound max gross weight loaded	Stainless steel, depleted uranium shield, hydrogenous neutron shield	Light Weight Truck	
TN-681019293	Transnuclear	68 BWR fuel assemblies	272,000 pounds maximum weight of package	Steel cask, gamma shield of steel, borated polyester resin compound cast in aluminum neutron shield	Rail	
TN-91019016	Transnuclear	7 BWR assemblies	38,110 kgmax load	Lead, steel, and resin shield	Truck	

Source: U.S. DOE RAMPAC Certification Web site, http://rampac.energy.gov/RAMPAC_Home.htm. Accessed April 2008.

Each cask used for transportation of used nuclear fuel or high-level waste must have a certificate of compliance issued by the NRC in accordance with 10 CFR 71, "Packaging and Transportation of Radioactive Material." The application for that certificate must contain a package description (10 CFR 71.33), a package evaluation (10 CFR 71.35), and a quality assurance program description (10 CFR 71.37). The description of the cask needs to be sufficient to allow the NRC to evaluate whether the cask meets all of the requirements. It includes such information as construction materials, size, and weight of the cask, fabrication methods, limits on the composition of the cargo the cask may carry, and means for heat transfer and dissipation. The described package must undergo evaluations described in detail in subparts E and F of the code. It must also meet the quality assurance requirements in subpart H of the code. The NRC has issued a document (NUREG-1617) titled "Standard Review Plan for Transportation Packages for Spent Nuclear Fuel" that provides detailed guidance to any company wishing to obtain a certificate of compliance.

Before issuing a certificate of compliance to a transportation cask for used nuclear fuel, the NRC must be convinced that the cask will confine the radioactive material and meet all other regulatory requirements under normal and a set of hypothetical accident conditions. The regulations say the cask "must be evaluated by subjecting a specimen or scale model to a specific test, or by another method of demonstration acceptable to the Commission, as appropriate for the particular feature being considered" (10 CFR 71.41(a)). Although computer simulation of a test may technically be allowed, typically the cask is subjected to a series of tests to determine whether it meets the requirements under four hypothetical accident conditions. Four tests representing accident conditions are conducted in sequence on a single cask specimen:

1. *Drop test*. Drop the cask from 30 feet onto a hard, unyielding surface in an orientation most likely to damage the cask.
2. *Puncture test*. Drop the cask from 40 inches on a 6-inch diameter shaft in an orientation most likely to result in damage.
3. *Fire test*. Engulf cask fully in a fire at 1475°F for 30 minutes.
4. *Immersion test*. Place cask under 3 feet of water for 30 minutes.

Figure 8.8 illustrates the sequence of tests. There is a fifth test, but it is conducted on a fresh cask rather than one that has been through the first four tests. The undamaged cask is immersed in water and subjected to pressure equivalent to water 50 feet deep (10 CFR 71.73).

7 TRANSPORTATION OF OTHER RADIOACTIVE MATERIALS

Although this chapter has focused on transportation of used nuclear fuel and high-level waste, highly radioactive materials that require special care, many other forms of radioactive material are routinely shipped in the United States.

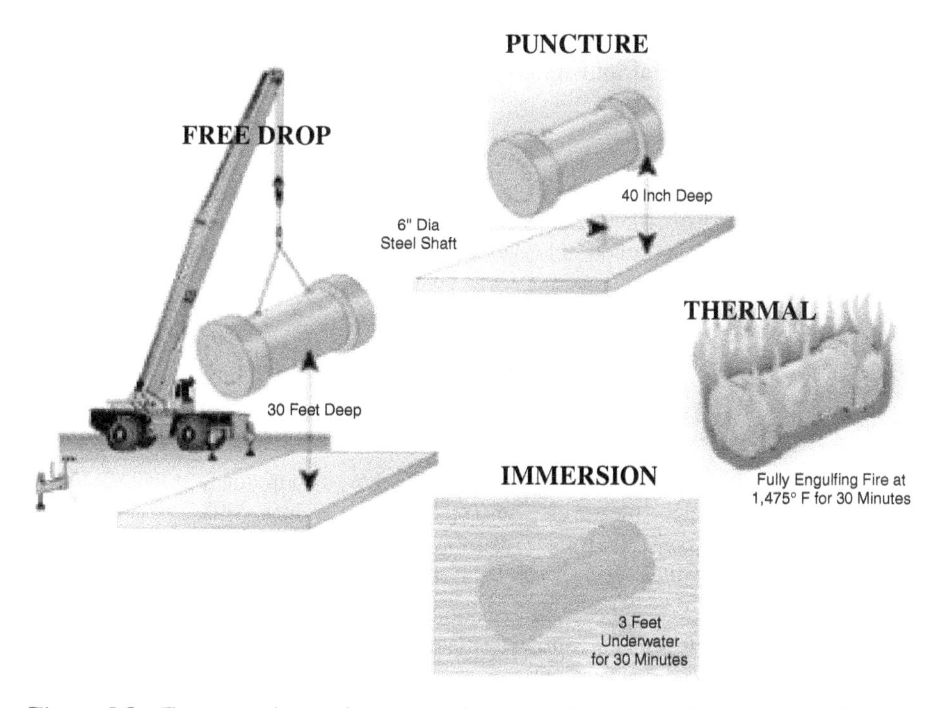

Figure 8.8 Transportation casks test requirements (Courtesy of Department of Energy)

According to the NRC, "about 3 million packages of radioactive materials are shipped each year in the United States, either by highway, rail, air, or water" (USNRC 2008b). These shipments include isotopes for use in diagnosing and treating diseases and small amounts of radioactive material for use in research. For example, researchers in agriculture, biology, chemistry, medicine, and many other fields use tiny amounts of radioactive material as tracers in their research. Many radioactive materials are used for industrial purposes such as nondestructive testing of welds, gages to help control thickness of paper or metal during manufacturing, and components of some smoke detectors. Larger amounts of radioactive material are used to sterilize medical supplies or to kill insects, bacteria, or mold in some foods and spices. Most of these uses of radioactive materials generate small amounts of contaminated waste, typically called low-level radioactive waste, that must be shipped to a designated disposal facility. Shipments of all of these types of radioactive materials are jointly regulated by the NRC through 10 CFR 71, and the Department of Transportation through 49 CFR 107, 171–180, and 390–397.

Acknowledgments

Three nuclear engineering students at Purdue University were instrumental in gathering and organizing the information presented in this chapter. They were

Sheila Bolbolan, Kevin Chesterfield, and Teandra Pfeil. Their contributions were substantial and much appreciated.

REFERENCES

Bradish, David. 2005. *Dollars and the nuclear waste fund*. NEI Nuclear Notes. http://neinuclearnotes.blogspot.com/2005/06/dollars-and-nuclear-waste-fund.html. Retrieved June 2008.

Fentiman, A. W., J. A. Henkel, and R. J. Veley. 1996. *How are people protected from ionizing radiation?* Columbus, OH: Ohio State University. http://ohioline.osu.edu/~rer/rerhtml/rer_26.html. Retrieved June 2008.

Global Nuclear Energy Partnership. 2007. *Global Nuclear Energy Partnership strategic plan* (GNEP-167312, Rev. 0).

Ryskamp, John M. 2003. Nuclear fuel cycle. http://nuclear.inel.gov/docs/papers-presentations/nuclear_fuel_cycle_3-5-03.pdf. Retrieved June 2008.

Shappert, L.B. 1998. Oak Ridge National Laboratory. *The Radioactive Materials Packaging Handbook*(ORNL/M-5003). Oak Ridge, Tennessee 47831-6285.

U.S. Code of Federal Regulations (CFR). 49 CFR Part 172. http://ecfr.gpoaccess.gov/cgi/t/text/text-idx?c=ecfrbrowse/Title49/49cfr172_main_02tpl. Retrieved June 2008.

U.S. Code of Federal Regulations. 10 CFR 71. http://www.nrc.gov/reading-rm/doc-collections/cfr/part071. Retrieved on June 2008.

U.S. Department of Energy. 2008a. *Central Internet Database*. http://cid.em.doe.gov/. Retrieved June 2008.

U.S. Department of Energy. 2008b. *U.S. Department of Energy Awards Contracts for Waste Storage Canisters for Yucca Mountain*. Press release, May 21.

U.S. Department of Energy. 2008c. *The Department of Energy's Website for information on radioactive material packaging*. Certificate retrieval page. http://rampac.energy.gov/RAMPAC_Home.htm. Retrieved April 2008.

U.S. Department of Energy. 2008d. *How much nuclear waste is in the United States?* http://www.ocrwm.doe.gov/ym_repository/about_project/waste_explained/howmuch.shtml. Retrieved June 2008.

U.S. Department of Energy. 2008e. *West Valley Demonstration Project*. http://www.wv.doe.gov.

U.S. Department of Energy: Energy Information Agency. 2004. *Spent nuclear fuel*. http://www.eia.doe.gov/cneaf/nuclear/spent_fuel/ussnfdata.html. Retrieved June 2008.

U.S. Department of Energy: Office of Civilian Radioactive Waste Management. 2003. *Strategic plan for the safe transportation of spent nuclear fuel and high-level radioactive waste to Yucca Mountain: A guide to stakeholder interactions*. http://www.ocrwm.doe.gov/transport/pdf/tsp.pdf. Accessed June 2008.

U.S. Department of Energy: Office of Civilian Radioactive Waste Management. 2004. *Nuclear waste policy act*. Washington, D.C: U.S. Government Printing Office.

U.S. Department of Energy: Office of Civilian Radioactive Waste Management. 2008. *Department of Energy policy statement for use of dedicated trains for waste shipments to Yucca Mountain*. http://www.ocrwm.doe.gov/transport/pdf/dts_policy.pdf. Retrieved June 2008.

U.S. Department of Energy: Office of Nuclear Energy. 2008. *Recycling spent nuclear fuel*. Available at www.nuclear.energy.gov and www.gnep.gov.

U.S. Nuclear Regulatory Commission. 2000. NUREG-1617, Standard Review Plan for Transportation Packages for Spent Nuclear Fuel. Washington, DC: U.S. Nuclear Regulatory Commission.

U.S. Nuclear Regulatory Commission. 2007a. *Shipping requirements*. http://www.nrc.gov/materials/transportation/shipping.html. Retrieved June 2008.

U.S. Nuclear Regulatory Commission. 2007b. *Nuclear materials transportation package certification*. http://www.nrc.gov/materials/transportation/certification.html. Retrieved June 2008.

U.S. Nuclear Regulatory Commission. 2008a. *Find operating nuclear power reactors by location or name*. http://www.nrc.gov/info-finder/reactor/USMap Retrieved June 2008.

U.S. Nuclear Regulatory Commission. 2008b. *Nuclear materials transportation*. Retrieved June 2008, from http://www.nrc.gov/materials/transportation.html.

U.S. Nuclear Regulatory Commission. 2008c. *Typical spent fuel transportation casks*. Retrieved June 2008, from http://www.nrc.gov/waste/spent-fuel-storage/diagram-typical-trans-cask-system.doc.

World Nuclear Association. 2008. *Supply of uranium*. http://world-nuclear.org/info/inf75.html. Retrieved June 2008.

Yucca Mountain application filed. 2008. *C&E News* (June 9): 32.

CHAPTER 9

PIPE SYSTEM HYDRAULICS

Blake P. Tullis
Utah State University
Logan, Utah

1 INTRODUCTION

There are many factors that should be considered when designing a pipeline. They include (but are not limited to): discharge capacity (present and future); pipe diameter; maximum positive and negative steady state and transient pressures; pipe material and pressure class; joint type, fluid type, and soil type; pumping requirements; the need for storage in the system; live and static external loads; the use of proper bedding material, installation techniques, and air valves to reduce the risk of pipe collapse due to negative internal pressures; control valve selection and operation; and proper filling, flushing, and draining protocols.

Negative pressures in the pipeline can cause external materials to leach into the pipe through joints, potentially contaminating the fluid. As a worst-case scenario, poor pipe design could lead to the collapse of the pipe. When the pipeline carries a fluid that may be hazardous if introduced to the local environment, additional precautions or design considerations are warranted. Such considerations should include leak detection monitoring, isolation valves distributed along the length

Table 9.1 Nomenclature

Symbol	Meaning
$\dot{m}$	Mass flow rate
ρ	Density
γ	Specific weight
μ	Dynamic viscosity
υ	Kinematic viscosity
A	Area
A_m	Area at inlet to pipe fitting for minor losses
A_p	Area of pipe
C	Total system loss coefficient, Equation 13
D	Diameter
e	Pipe roughness
F	Momentum force
F	Friction factor
G	Acceleration of gravity
H_f	Total head loss, friction and minor
H_L	System head loss
H_m	Head loss due to minor losses
H_p	Pump head
H_t	Turbine head
K	Minor loss coefficient
L	Pipe length
P	Pressure
Q	Flow rate
Re	Reynolds number
S	Second (time)
V	Velocity
Z	Elevation

of the pipeline to limit the volume of potential leaks, and the use of noncorrosive pipe and seal materials.

This chapter examines the factors in designing the proper pipe system. The terms in Table 9.1 will be referenced throughout the chapter. The equations presented in section 1.1 and 1.2 are important for proper pipe system design.

1.1 Basic Fluid Mechanics Equations

Solving fluid flow problems involves the application of one or more of the three basic equations: continuity, momentum, and energy. These three basic equations were developed from the law of conservation of mass, Newton's second law of motion, and the first law of thermodynamics.

The simplest form of the continuity equation is for one-dimensional, steady flow in a closed conduit. Applying continuity between any two sections gives:

$$\dot{m} = \rho_1 A_{p1} V_1 = \rho_2 A_{p2} V_2 \tag{1}$$

The continuity equation for incompressible fluids further simplifies to:

$$Q = A_{p1} V_1 = A_{p2} V_2 \tag{2}$$

A_p is the pipe cross sectional area, V is the mean velocity, Q is the volumetric flow rate, ρ is the fluid density, $\dot{m}$ is the mass flow rate, and the subscripts represent different locations in the pipe. Equations (1) and (2) are valid for steady flow in any rigid conduit so long as there is no addition or loss of liquid between sections 1 and 2.

For steady state pipe flow, the momentum equation (3) equates the change in momentum flux to the sum of forces acting on a control volume in a given direction (e.g., x direction) as shown in Figure 9.1. The most common forces are pressure, friction, and boundary forces.

$$\sum F_x = \rho_2 A_{p2} V_2^2 - \rho_1 A_{p1} V_1^2 \tag{3}$$

For incompressible flow, the momentum equation reduces to:

$$\sum F_x = \rho Q \left(V_2 - V_1 \right) \tag{4}$$

These equations are typically applied to three-dimensional, steady-state flow problems by adding equations in the y and z directions.

The Bernoulli equation, which is a steady-state energy equation, applied to incompressible pipe flow becomes:

$$\frac{P_1}{\gamma} + Z_1 + \frac{V_1^2}{2g} + H_p - H_t - H_L = \frac{P_2}{\gamma} + Z_2 + \frac{V_2^2}{2g} \tag{5}$$

The units in equation (5) are energy per unit weight of liquid (ft-lb/lb or N-m/N), commonly referred to as *head* and expressed in units of length. P/γ is the pressure head, Z is the elevation head (above a reference datum), and $V^2/(2g)$ is the velocity head. H_p is the total dynamic head added by the pump, H_t is the total dynamic head removed by a turbine, and H_L is the head lost to friction and minor losses. The subscripts represent different locations in the system, as shown in Figure 9.2.

1.2 Energy Losses

The balance between the driving head available to the system and the energy losses determines the flow rate through a pipe system or the head-discharge

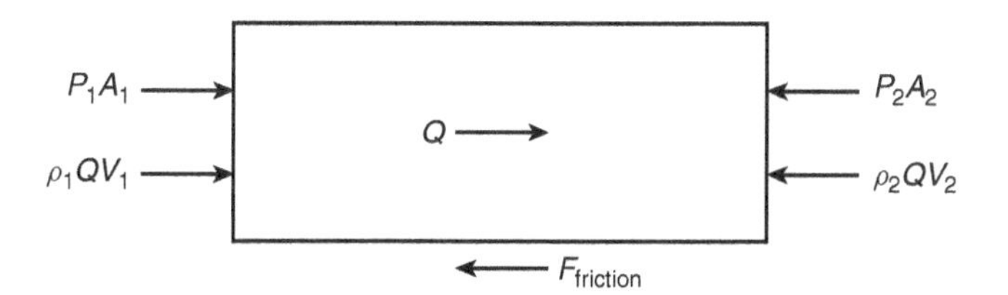

Figure 9.1 Control volume and momentum equation components

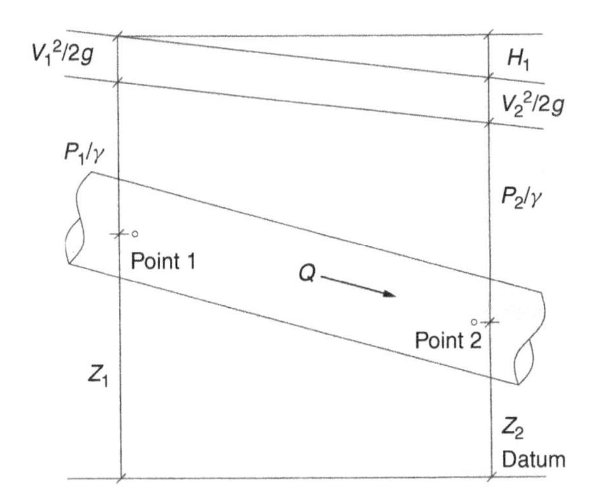

Figure 9.2 Schematic illustrating energy equation components (excluding H_t and H_p components)

relationship. For a pipe system that conveys flow between two large reservoirs, the driving head is the difference between the upstream and downstream reservoir elevations plus any additional head added by pumps (H_p) if present. System energy losses (H_L) include friction loss (H_f) caused by boundary shear stresses and minor losses (H_m) resulting from the dissipation of turbulent eddies created by flow passing through pipe fittings (e.g., elbows, tees, expansion, contractions, etc.), valves, and pipe inlets and outlets.

$$H_L = H_f + H_m \tag{6}$$

2 FLUID FRICTION

The energy loss associated with boundary shear forces on incompressible flow in closed conduits is referred to as *friction loss*. The following discussion is developed for circular pipe; however, the results can be applied to noncircular pipe by replacing the pipe diameter with four times the hydraulic radius ($4R_h$). R_h is the conduit cross-sectional area (A) divided by the conduit circumference. The analysis in this section can also be applied to gases and vapors, provided that the Mach number in the duct does not exceed 0.3. For Mach numbers greater than 0.3, the compressibility effect becomes significant and additional considerations are required.

H_f is dependent upon pipe diameter (D), pipe length (L), pipe wall material roughness (k_s), fluid density (ρ) or specific weight (γ), fluid kinematic viscosity (υ), and mean flow velocity (V). Dimensional analysis can be used to provide a functional relationship between the H_f, pipe dimensions, fluid properties, and flow parameters. The resulting equations are called the Darcy-Weisbach equation

(7) and the Reynolds number (8).

$$H_f = \frac{fLV^2}{D2g} = \frac{fLQ^2}{D2gA_p^2} \tag{7}$$

$$\mathrm{Re} = \frac{\rho VD}{\mu} = \frac{VD}{\upsilon} \tag{8}$$

In equations (7) and (8), Q is the volumetric flow rate, f is the friction factor that characterizes the flow resistance, and μ dynamic viscosity of the fluid. Experimental values of f have been determined for numerous pipes of varied geometry and material; the results of which were used to create the Moody diagram shown as Figure 9.3. The Moody diagram is a graphical representation of the Colebrook-White equation (9).

$$\frac{1}{\sqrt{f}} = 1.74 - 0.869 \ln\left(\frac{2k_s}{D} + \frac{18.7}{\mathrm{Re}\sqrt{\mathrm{f}}}\right) \tag{9}$$

Solving the Colebrook-White equation for f requires a trial-and-error or iterative solution method. Swamee and Jain (1976) developed an explicit relationship that approximates the Colebrook-White equation and Moody diagram.

$$f = \frac{0.25}{\left[\log\left(\frac{k_s}{3.7D} + \frac{5.74}{\mathrm{Re}^{0.9}}\right)\right]^2} \tag{10}$$

The friction factor and the subsequent friction loss of a flowing liquid depend on whether the flow is laminar or turbulent. *Laminar flow* exists when viscous forces are large compared to inertial forces. When inertial forces are large compared to viscous forces, the flow is considered turbulent. The Reynolds number (Re) is the ratio of inertia forces to viscous forces and is a convenient parameter for characterizing laminar and turbulent flow. For Re $< 2{,}000$, the flow is laminar and f is solely dependent on the Re ($f = 64/\mathrm{Re}$). Most practical pipe flow problems are in the turbulent region. The velocity of water flowing in a 1 m diameter pipe at 20°C would have to be $\leq$ 2 mm/sec to be in the laminar range.

When $2{,}000 < \mathrm{Re} < 4{,}000$, the flow is transitioning from laminar flow to turbulent flow and is unstable (critical zone in Figure 9.3). In this range, friction loss calculations are difficult because it is impossible to determine a unique value of f. Fortunately, most practical pipe flow problems involve Re $> 4{,}000$.

When Re $> 4{,}000$, the flow becomes turbulent and f is a function of both Re and the relative pipe roughness (k_s/D). k_s represents the average height of the material roughness elements of the conduit boundary. As the level of flow turbulence increases (increasing Re), a wholly rough turbulent condition, as shown in Figure 9.3, is eventually reached where f is only dependent upon k_s/D.

Figure 9.3 gives values of k_s for common pipe materials. Note that a range of k_s values, rather than one single value, is given for each material. This is because

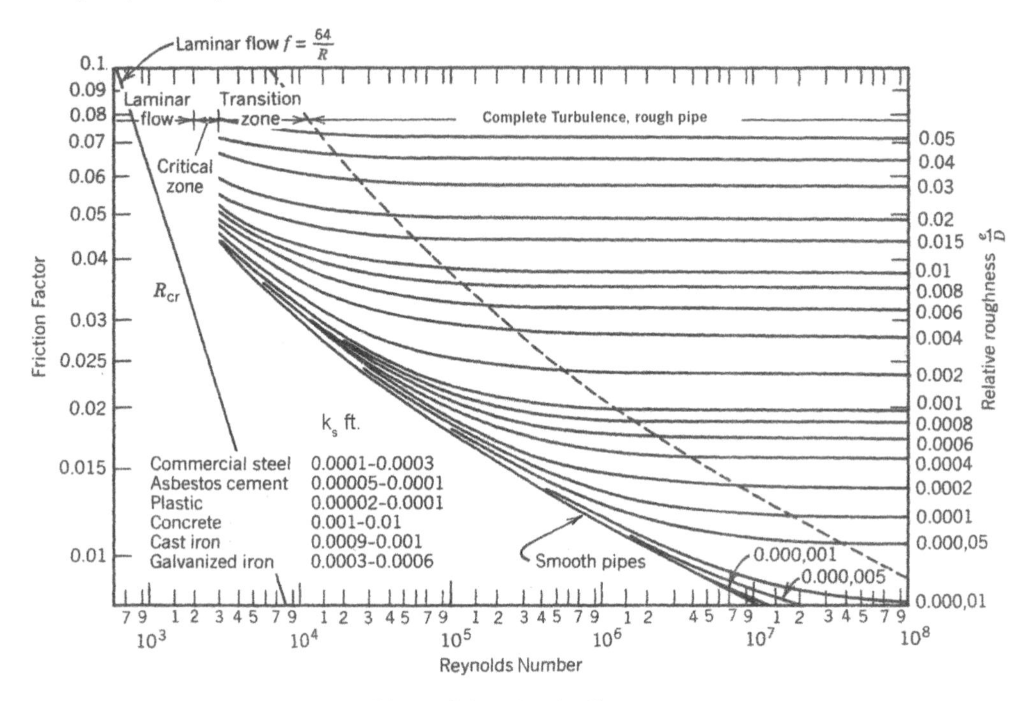

Figure 9.3 Moody diagram

pipe wall roughness typically varies during the life of the pipe as a result of pipe wall material removal and deposition, mineral deposition, or growth of organic matter. Manufacturing methods also cause variations in the surface roughness. The representative values of k_s listed in Figure 9.4 should be considered approximate (consult pipe manufacturer) and proper allowance should be made for these uncertainties.

To determine f using the Moody diagram (Figure 9.3), Re and k_s/D must be known. Re can be calculated directly if the water temperature, Q, and D are known. The problem then becomes obtaining a good value for k_s. If either the Q or D is not known, the solution to the Darcy-Weisbach or Swamee-Jain equations require a trial-and-error or iterative solution.

For long gravity-flow pipelines, the criterion for pipe diameter selection is simply finding the smallest pipe that can pass the required flow rate without the friction and minor losses exceeding the available head. The required flow rate for design should consider both current and future demand considerations. For pumped systems, optimizing the pipe diameter is based on an economic analysis that compares the installed pipe cost with the cost of building and operating the pumping plant. Pipe cost is proportional to D and friction loss, which is related to the required pumping head and operating cost, is inversely proportional to D. The optimum pipe diameter is selected as the one that provides the lowest total cost.

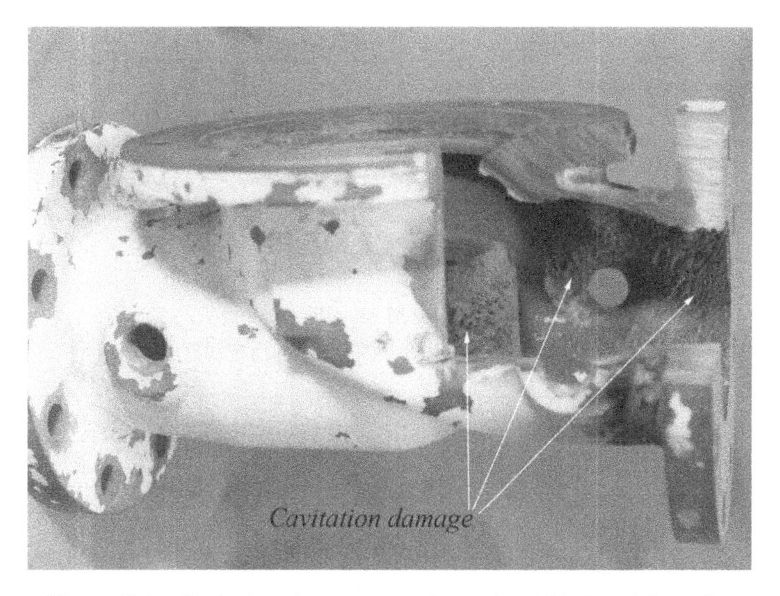

Figure 9.4 Cavitation damage near the outlet side of a globe valve

3 MINOR LOSSES

Flow passing through valves, orifices, elbows, transitions, and other pipe fittings generates an increase in turbulent eddies. Head loss associated with the dissipation of the turbulent eddies is referred to as minor loss. The term *minor loss* can be somewhat of a misnomer in some cases such as short piping systems where the friction loss is relatively small and the minor losses are responsible for the majority of the system head loss. Minor losses are represented as a loss coefficient (K) multiplied by the appropriate pipe velocity head:

$$H_m = K\frac{V^2}{2g} = K\frac{Q^2}{2gA_m^2} \tag{11}$$

In equation (11), A_m is the pipe flow area at the inlet of the minor loss element.

Minor loss coefficients are generally available in most hydraulic engineering reference books (Miller 1990). Some minor loss coefficient data are presented in Table 9.2. The loss coefficients for fittings vary with material, manufacturer's method of fabrication, and installation. Minor loss coefficients for tees also vary with the flow direction and percent flow distribution between the branches.

The total system head loss (H_l) can be calculated using the following equation provided the appropriate friction and minor loss coefficients are used.

$$H_L = \left(\sum \frac{fL}{D2gA_p^2} + \sum \frac{K}{2gA_m^2}\right) Q^2 = CQ^2 \tag{12}$$

Table 9.2 Minor Loss Coefficients

Minor Loss Element	K
Pipe Inlets	
Projecting pipe	0.5 to 0.9
Sharp corner-flush	0.50
Slightly rounded	0.04 to 0.5
Bell mouth	0.03 to 0.1
Sudden expansions[a] (based on inlet velocity, V_1)	$(1 - A_{p1}/A_{p2})^2$
Sudden contractions[b] (based on outlet velocity, V_2)	$(1/C_c - 1)^2$

A_{p2}/A_{p1}	0.1	0.2	0.3	0.4	0.5	0.6	0.7	0.8	0.9
C_c	0.624	0.632	0.643	0.659	0.681	0.712	0.755	0.813	0.892

Minor Loss Element	K
Steel Bell Reducers[c] (Welded, $D_1/D_2 = 1.2$ to 1.33)	0.053 to 0.23
Steel Bell Expanders[c] (Welded, $D_2/D_1 = 1.2$ to 1.33)	0.02 to 0.11
PVC Fabricated Reducers[c] ($D_1/D_2 = 1.33$ to 1.5)	0.12 to 0.68
PVC Fabricated Expanders[c] ($D_2/D_1 = 1.2$ to 1.33)	0.07 to 1.19
Bends[d]	
Short radius, curve radius/pipe diameter = 1	
90°	0.3 to 0.6
45°	0.10
30°	0.06
Long radius, curve radius/pipe diameter = 1.5	
90°	0.07 to 0.33
45°	0.09
30°	0.06
Mitered (one miter)	
90°	1.10
60°	0.40 to 0.59
45°	0.35 to 0.44
30°	0.11 to 0.19

Valves (full-open)	Average K Values	Range
Check valves[e]		
Swing check	1.0	0.29 to 2.2
Tilt disc	1.2	0.27 to 2.62
Lift	4.6	0.85 to 9.1
Double-door	1.32	1.0 to 1.8
Gate	0.15	0.1 to 0.3
Butterfly	0.56	0.2 to 0.6
Globe	4.0	3 to 10

[a] See Streeter and Wylie (1975)
[b] See Streeter and Wylie (1975)
[c] See Rahmeyer (2002)
[d] See Miller (1990)
[e] See Kalsi and Tullis (1993)

The summation symbols suggest that multiple minor loss elements and pipe sections of differing size and/or roughness may exist in a pipe system. It is important to use the actual pipe inside diameter and corresponding cross-sectional area of each pipe section and minor loss element. For a given nominal pipe size, the inside diameter will vary with pressure class. Pipe head-discharge relationships are developed by replacing the H_L term in equation (5) with equation (12)

4 PIPE SELECTION

Materials commonly used for pressure pipe transporting liquids are ductile iron, concrete, steel, fiberglass, PVC, and high-density polyethylene (HDPE). National committees have developed specifications for most types of pipe products. The specifications discuss external loads, internal design pressure, available sizes, quality of materials, corrosive environments, installation practices, and linings. Standards are available from the following organizations:

- American Water Works Association (AWWA)
- American Society of Mechanical Engineers (ASME)
- American Society for Testing and Materials (ASTM)
- American National Standards Institute (ANSI)
- Plastic Pipe Institute (PPI)
- Federal Specifications (FED)
- Canadian Standards Association (CSA)

In addition, manuals and other standards have been published by various manufacturers and manufacturing associations. These specifications and standards should guide in the selection of pipe material. ASCE (1992) contains a description of most of these pipe materials and a list of the specifications from the various organizations associated with each material. The document also discusses the various pipe linings available for corrosion protection. There are several relevant publications for selecting the proper type and pressure class of pipe (ASCE 1992; ASCE 1993; AWWA 1995; AWWA 1989; AWWA 2003; and PPI 1980).

For low-pressure liquid applications, available pipe materials include un-reinforced concrete; corrugated, spiral-ribbed, and smooth sheet metal; and HDPE (high-density polyethylene). The choice of a material for a given application depends on pipe size, pressure requirements, resistance to collapse from internal vacuum pressures and/or external loads, resistance to internal and external corrosion, ease of handling and installing, useful life, economics, and the available driving head versus the friction loss characteristics of the pipe.

When dealing with corrosive liquids or gasses, the corrosion-resistant nature of the sealing material in the pipe joints must also be taken into consideration. Most seals are provided by a flat rubber gasket sandwiched between two flanges or an O-ring that fits inside the bell-end of one pipe and the spigot-end of the adjoining pipe. Proper selection of pipe and seal materials will reduce the risk of leaks and possible contamination of the surrounding environment.

5 VALVES

The minor loss coefficients for a variety of different valve types (in the full-open position) are included in Table 9.2. Valves differ from other minor loss elements in that their loss coefficients and corresponding head loss are variable with valve opening. There are more things to be considered regarding valves than loss coefficients, however. Valves serve a variety of functions in piping systems, such as flow isolation, flow rate control, pressure regulation, energy dissipation, reverse flow prevention, and releasing/admitting air. This section discusses characteristics of and principles for selecting and operating control valves, check valves, air valves, and pressure relief valves. For a description of various types of valves and details regarding their function, see Tullis and Tullis (2005).

5.1 Control Valves

The function of a control valve is to regulate the flow rate through the piping system. In general, as the valve opening increases, the loss coefficient decreases and the flow rate increases. The efficiency with which a specific valve controls the flow rate is system dependent and is related to the ratio of the valve loss to the rest of the system losses. In a long pipe system where the friction loss is much greater than the valve minor loss (over a significant portion of the valve stroke), the valve will only control flow at small valve openings. The same valve, installed in a short, low-friction system, will have much improved flow control characteristics. For more information regarding control valves, see Tullis (1989, 1993).

5.1.1 Cavitation

Cavitation is one possible negative side effect that can result from head loss across control valves or other minor losses. Cavitation is the process of converting water to vapor when the local pressure in a section of pipe reaches vapor pressure. As flow passes through valves or other minor loss elements, the flow is typically accelerated and concentrated as a result of flow separation from the boundary. Flow separation typically results from flow path obstructions and abrupt changes in flow direction. As a result, the flow velocity increases and the pressure decreases (per the Bernoulli relationship). The turbulent eddies produced in such flow patterns generate energy dissipation as well as localized low pressures associated with the large eddy rotational velocities. When the local pressure inside the eddy reaches vapor pressure, vapor bubbles form. As the eddies dissipate, the pressure increases (i.e., pressure recovery) and the vapor bubbles become unstable and collapse. Low levels of cavitation will generate a light, crackling noise. High levels of cavitation can generate tremendous levels of noise and vibration, and can cause damage to the valve, fitting, or pipe wall through material removal. An example of cavitation damage in a globe valve is shown in Figure 9.4.

Different valves have different cavitation characteristics. Different applications also warrant different levels of cavitation restrictions. A control valve that experiences continual use should be operated below the cavitation damage threshold. A pressure-relief valve, whose purpose is to keep line pressures below a maximum level for process or safety reasons, is typically required to operate at the maximum discharge possible. For this type of application, pressure relief typically supersedes valve cavitation damage concerns, and/or the valve operation events are so infrequent that no appreciable damage will occur over the life of the valve. Additional information on valve cavitation is presented by Tullis (1989, 1993).

5.1.2 Hydraulic Transients

Most pipe analyses correspond to steady-state flow conditions. When a valve opening position is changed, a pump is started or shut down, or a pipe rupture occurs, an unsteady pipe flow condition is generated in the pipe system. Sudden flow decelerations or accelerations can produce traveling transient pressure waves with magnitudes significantly higher or lower then the steady-state line pressures. The duration of transient pressure waves are typically short-lived. The resulting damage, however, can be severe if the pipe is not designed for the extreme positive and negative pressures associated with transient events. As an example, if the flow velocity in a steel pipe were decreased instantly (or over a very short period of time) by 1 m/s, transient pressure head increase of approximately 100 m would result. Every pipe system should have at least a cursory transient analysis performed to identify the possibility of serious transients and decide whether a detailed analysis is necessary. If an analysis indicates that transients are a problem, there are at least five methods of controlling transients:

1. Increase the closing time of the control valve.
2. Use a smaller valve to provide better control.
3. Design special facilities for filling, flushing, and removing air from pipelines.
4. Increase the pressure class of the pipeline.
5. Use pressure relief valves, surge tanks, air chambers, and so on.

For more details regarding hydraulic transients, see Tullis (1989, 1993) and Wylie and Streeter (1993).

5.1.3 Torque

Selection of the correct operator and shaft size for a specific quarter-turn valve and application requires knowing the maximum torque that the valve will experience during operation. This requires analyzing the system for the entire range of expected flow conditions in order to identify the maximum flow torque condition. There are four primary sources of torques for quarter-turn valves:

1. *Seating torque*. Torque is created by the disk moving into or out of the seat (sealing surface).
2. *Bearing friction torque*. Torque is required to overcome the frictional resistance that forms between the valve shaft and its bearing surfaces in the valve body.
3. *Packing friction torque*. Torque is required to overcome the friction resistance that forms between the valve shaft and the packing material used to prevent leaks between the shaft and valve body.
4. *Hydrodynamic (flow) torque*. Torque is created by the flow momentum and differential pressure across the disk.

These torque values are usually determined experimentally and should be available from the valve manufacturer. All four torques should be evaluated to determine the maximum torque for a given valve. Operating torque is normally greater when the valve is being opened because the hydrodynamic, bearing, and packing torques all oppose the direction of valve motion. During valve closure, the bearing and packing torques act in the direction opposite to that of the hydrodynamic torque. The seating or unseating torque can also be significant. [I think this works]. It is the responsibility of the valve manufacturer to provide the flow and torque characteristics and limits for their valve. It is the responsibility of the system engineers and operators to see that the valves are operated within these limits.

5.1.4 Restricted Operating Range of Valve Openings

Many conventional control valves warrant a restricted operating range of valve openings, as they cannot safely and/or accurately regulate flow near the closed and full-open positions. Near the closed position, two potential problems include seat (sealing surface) damage due to high velocities and inability to accurately set the valve opening when the valve operator experiences hysteretic effects in its valve positioning. Near full open, some valves lose control, meaning there is no change in flow rate with valve position change. For globe-style valves, this occurs when the stroke is too long. The full-open characteristics of butterfly valves are influenced by the shape of the disc and changes in the flow pattern around the disc at large openings. For some disc shapes, the flow can actually decrease as the valve disc position moves between 90 and 100 percent open. This problem is magnified when the valve is installed in a long system where the valve loss is small compared to the system friction loss. Some quarter-turn valves experience a torque-reversal near the full-open position, which can cause the disk to flutter and which increases the potential for accelerated wear.

In summary, when analyzing a flow control valve, six criteria should be considered:

1. The valve should not produce excessive pressure drop when full open.
2. The valve should control over about 50 percent of its movement (i.e., closing the valve half way should reduce the flow rate by 10%).
3. The cavitation intensity should be limited to the appropriate level.
4. The valve should be operated so that the resulting transient pressures do not exceed the safe limits of the system.
5. The valve operator and shaft for quarter-turn valves should be compatible with the valve maximum torque requirement.
6. Restrictions in the range of valve-opening operation should be identified. Some valves should not be operated near the closed or fully open positions.

5.2 Check Valves

Check valves are used to prevent reverse flows resulting from pump shutdown, for example, or pipe rupture. Check valves are designed to pass flow efficiently in the forward-flow direction and to close quickly when the flow reverses. Most check valves are swing-type, spring-actuated, or a combination of both. When selecting a check valve, the characteristics of the valve should be compatible with the characteristics or requirements of the system. Selecting the wrong type or size of check valve can result in poor performance, severe transients, and frequent repairs (Kalsi and Tullis 1993). For a description of the characteristics of common types of check valves, see Kalsi and Tullis (1993) and Tullis and Tullis (2005). Five characteristics of check valves should be considered in the selection process:

1. Closure speed of check valves relative to the rate of flow reversal of the system
2. Stability of the disk and its sensitivity to local turbulence effects
3. The flow rate required to fully open and firmly back seat the disc
4. The head loss at the maximum and typical flow rates
5. Sealing effectiveness and ease of maintenance

It is a mistake to oversize a swing check valve located just downstream from a disturbance such as a pump, elbow, or control valve. The disk will not firmly back seat, and it will be subjected to severe motion and accelerate wear. All check valves have components that experience wear. If the valves are not inspected and maintained periodically, these valve components may eventually fail, rendering the check valve ineffective. Operating check valves with the disk firmly seated in the full-open position during normal operation will reduce the amount and rate of component wear and will increase the useful life of the check valve.

It is also important to consider the protection against harmful transient pressure wave generation by check valve operation. Transient pressures are maximized in a system with a fast flow reversal and a slow moving check valve (i.e., swing-check valves). From a limiting transient pressure standpoint, a fast acting, spring-actuated check valve that closes at about the same time as the flow comes to rest (just before reversing direction) would be ideal. Examples of fast-closing, spring-actuated check valves included nozzle, silent, duo, double-door, and lift check valves.

5.3 Air Valves

Air valves release unwanted air volumes that collect in a pipe system (typically at high points) and admit air into the pipe when the pressure becomes sub-atmospheric. Negative pressures can occur during normal operation if the hydraulic grade line or piezometric head (elevation plus pressure head) falls below the pipeline profile, such as at a high point. Intentional or unintentional (pipe rupture) draining of the pipeline can also result in negative local pressures. Large, moving air pockets generated through the pipe-filling process, unsubmerged pipe inlets, or air admitted by air valves can result in the generation of dangerous transient pressures when passing through flow restrictions such as control valves.

5.4 Air/Vacuum Valves

There are three types of automatic air valves: (1) air/vacuum valves, (2) air release valves, and (3) combination valves. The air/vacuum valve is designed for releasing large quantities of air while the pipe is being filled and admitting air when the pipe is being drained. Air/vacuum valves typically contain a float, which rises and closes a discharge port as the valve body fills with water. Once the line is pressurized, the float cannot reopen to remove air that may subsequently accumulate. If the pressure becomes negative during a transient or while draining, the float drops and admits air into the line. At least one of the air/vacuum valves in the system should be sized for the maximum discharge associated with a full pipe break. Air/vacuum valves, when sized based on vacuum service, should be sized to ensure protection against negative pressures caused by line breaks. Note that the pipe fill rate should be controlled by the water inflow rate rather than the discharge capacity of the air/vacuum valve(s) to avoid transients when the air/vacuum valve closes.

5.5 Air Release Valves

Air release valves contain a small orifice and are designed to release small quantities of pressurized air that are not released by the air/vacuum valves during the pipe filling or normal operation. The small orifice is controlled by a plunger activated by a float at the end of a lever arm. As air accumulates in the valve body, the float drops and opens the orifice. As the air is expelled, the float rises and closes the orifice.

5.6 Combination Valves

The combination valve includes both an air/vacuum valve and an air release valve. The installation can either consist of an air/vacuum valve and an air release valve plumbed in parallel, or the two can be housed in a single valve body. Most air valve installations require combination valves. Guidelines for sizing air valves are available from valve manufacturers and AWWA (2001). Large manual air release valves should not be used because they can be can cause severe transients.

5.7 Pressure Relief Valves

Pressure relief valves (PRVs) are installed to limit maximum system pressures. They can open automatically when the system pressure exceeds a set pressure, or they can be programmed to open in anticipation of a transient or pressure surge. There are two general types of relief valves: nonpilot and pilot actuated. The type chosen depends on the size of valve required and whether the valve opening or closing rates need to be controlled.

Nonpilot actuated PRVs have springs that hold the disk closed until the pipe pressure exceeds the spring setting. The disk opening is proportional to the overpressurization. They automatically close when the pressure drops below the spring setting. Nonpilot valves are fast acting with no valve opening or closing speed control. They are typically limited to smaller valve sizes.

Pilot actuated PRVs are typically globe valves, which are opened and closed using the upstream and/or downstream line pressure and a restoring spring. Throttling valves in the pilot system controls the PRV opening and closing speeds. Pilot valves typically actuate more slowly than nonpilot valves. The range of operating pressures can be adjusted by varying the pilot valve settings, and the pilot system can be programmed to operate when a pressure surge generated elsewhere in the system is anticipated. There are no size-specific limits to pilot activated PRVs.

6 CENTRIFUGAL PUMP SELECTION AND PERFORMANCE

When gravititational forces are insufficient to drive the required flow rate, pumps must be added to the system. Proper pump selection includes matching the pump capacity to the system requirements. Present and future flow requirements should be considered when selecting the design flow rate or range of design flow rates. The pump selection process requires developing an energy-based head-discharge relationship or system equation for the piping system. Eliminating the turbine component (H_t), the pump total dynamic head is defined by solving for H_p in equation (5).

$$H_p = \frac{(V_2^2 - V_1^2)}{2g} + \frac{(P_2 - P_1)}{\gamma} + Z_2 - Z_1 + H_L \tag{13}$$

For a pump supplying water between two reservoirs or tanks, the pump head required to produce a given discharge can be expressed as

$$H_p = \Delta Z + H_L \text{ or } \quad H_p = \Delta Z + CQ^2 \tag{14}$$

C is defined in equation (12), and ΔZ (i.e., $Z_2 - Z_1$) is referred to as the *static lift*. Figure 9.5 shows a system curve for a pipe having a static lift of 65 m and moderate friction losses. When the reservoir elevations are variable, a family of system curves will exist, describing the varying ΔZ values and corresponding flow rates.

Figure 9.5 shows three pump curves, which represent a single pump with three different impeller sizes. The intersection of the system and pump curves identifies the flow rate and the head (H_p) generated by the pump for each impeller size. Figure 9.5 also includes pump efficiency, cavitation (net positive suction head required, or NPSHr), and brake horsepower (bhp) information. For the system represented in Figure 9.5, impellers A and B would be a good choice based on efficiency. According to Figure 9.5, pump B will produce a head and flow rate of 93 m and 480 l/s for this system. NPSHr represents the minimum upstream total head, relative to absolute vapor pressure (datum) at the suction side of the pump required to avoid cavitation levels that will decrease the pump performance. bhp is the power added by the pump to the flowing water.

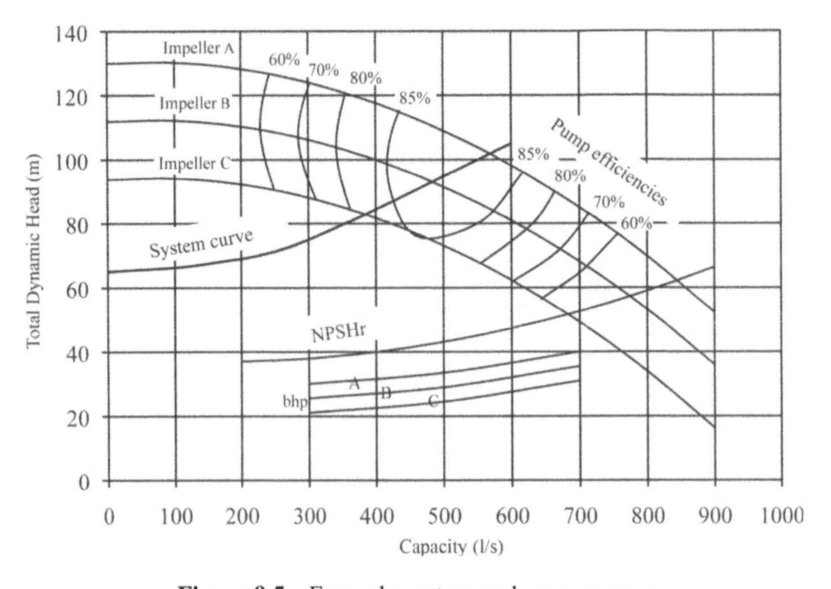

Figure 9.5 Example system and pump curves

Systems may require multiple pumps, placed either in parallel or in series. Parallel pumps are effective at increasing system discharge in low-friction, low-loss systems. Series pumps typically provide only modest increases in flow rate but significantly increase the head for high-friction or high-static lift systems.

When trying to match pumps to piping systems with variable head-discharge requirements, variable-speed pumps should be considered for installations when a wide range of flow conditions is required. The variable-speed drive can operate over a significant range of motor speeds. Variable-speed drives offer several advantages. Single-speed drives start the motor abruptly, subjecting the rotating element to high torque and electrical current surges several times the full load current, along with possible transient and vibration problems. In contrast, variable-speed drives provide a soft-start, gradually ramping the motor to operating speed.

One of the primary justifications for using a variable-speed drive is the cost savings resulting from reduced power demands and reduced maintenance of the motor, pump, and discharge control valve. With constant-speed pumps, flow reduction in a given system can only be achieved by throttling the control valve. This is somewhat analogous to operating an automobile at full throttle and trying to control speed by applying the brake. The excess pressure drop across the valve to reduce the flow rate results in wasted energy and creates the possibility of cavitation.

Additional details regarding series and parallel pump selection and performance, pump cavitation, pump operation, and suction piping considerations are provided by Tullis (1989, 1993).

7 OTHER PIPELINE OPERATION CONSIDERATIONS

There are many details that need consideration in pipeline design. In addition to the items already discussed, a few additional design considerations are briefly discussed.

7.1 External Loads

There are situations where the external load is the controlling factor in determining the risk of pipe collapse. The magnitude of the external load and the resistance of the pipe to collapse depend on numerous factors, including internal pressure, pipe diameter, pipe material and wall thickness, pipe deformation (deviation from a circular cross section), trench width, depth of cover, specific weight of the soil, degree of soil saturation, type of backfill material, method used to backfill, degree of compaction, and live loads. The cumulative effect of all these sources of external loading requires considerable study and analysis, beyond the scope of this chapter. There are no simple guidelines for evaluating external pipe loads; the reader is referred to (Watkins and Anderson 2000) and

(Spangler and Handy 1973) for details on how to perform calculations of earth loading.

7.2 Maximum and Minimum Flow Velocities

As a general guideline, maximum and minimum pipe flow velocities should be limited. When the flow velocity is too low, insufficient drag force is available to carry air bubbles that collect along the crown of the pipe or at high points to air release valves or the pipe exit for removal. If the air volumes are allowed to grow during low-velocity conditions, the large air volumes become mobile and may cause transient problems in the system when velocities increase. Also, suspended sediments carried in the pipe flow are more likely to settle out of suspension in the pipe when the velocities are low. Generally, flow velocities greater than 1 m/s (3 ft/s) are sufficient to move trapped air through the system and to keep most sediment suspended.

Problems associated with high velocities include abrasion or erosion of the pipe wall, valves, and fittings; cavitation at control valves and other minor loss elements; increased friction and minor losses (energy loss increases with the square of the velocity; decrease in efficiency of air removal at air release valves; increased hydrodynamic torque on control valves; and increased risk of hydraulic transients. Each of these should be considered before making the final pipe diameter selection. A typical upper velocity for many applications is 6 m/s (20 ft/s). With proper pipe design and analysis, however, higher velocities can be tolerated.

REFERENCES

ASCE. 1992. *Pressure pipeline design for water and wastewater*. Prepared by the Committee on Pipeline Planning of the Pipeline Division of the American Society of Civil Engineers.

ASCE. 1993. Steel penstocks. Prepared by the ASCE Task Committee on Manual of Practice for Steel Penstocks No. 79, Energy Division, American Society of Civil Engineers.

AWWA. 1989. *Steel pipe—a guide for design and installation* (M11). 1989. Denver, CO: American Water Works Association.

AWWA. 2003. *Ductile-iron pipe and fitting* (M41), 2nd ed. 2003. Denver, CO: American Water Works Association.

AWWA. 1995. *Concrete pressure pipe* (M9). Denver, CO: American Water Works Association.

AWWA. 2001. *Butterfly valves: Torque, head loss and cavitation analysis* (M49). Denver, CO: American Water Works Association.

Kalsi Engineering and Tullis Engineering Consultants. 1993. Application guide for check valves in nuclear power plants, Revision 1. NP-5479. Prepared for Nuclear Maintenance Applications Center, Charlotte, North Carolina.

Miller, D. S. 1990. *Internal flow systems—Design and performance prediction*, 2nd ed. Houston, TX: Gulf Publishing Company.

PPI. 1980. *PVC pipe design and installation* (M23). Plastics Pipe Institute, Inc.

Rahmeyer, W. 2002. Pressure loss coefficients of threaded and forged weld pipe fittings for ells, reducing ells, and pipe reducers. Technical paper H-1405. Atlanta, GA: American Society of Heating, Refrigeration and Air Conditioning Engineering.

Spangler, M. G., and R. L. Handy. 1973. *Soil engineering*, 3rd ed. New York: Intext Educational Publishers, chs. 25 and 26.

Streeter, V. L., and E. B. Wylie. 1975. *Fluid mechanics*, 6th ed. New York: McGraw-Hill, 752 pp.

Swamee, P. K., and A. K. Jain. 1976. Explicit equations for pipe-flow problems. *Jour. Hydraulic Div. ASCE* 102 (HY5).

Tullis, J. P. 1989. *Hydraulics of pipelines—Pumps, valves, cavitation, transients*, New York: Wiley.

Tullis, J. P. 1993. *Cavitation guide for control valves*, NUREG/CR-6031. Washington, DC: U.S. Nuclear Regulatory Commission.

Tullis, B. P., and J. P. Tullis. 2005. Valves. In *The engineering handbook*, ed. R. C. Dorf. Boca Raton, FL: CRC Press.

Watkins, R. K., and L. R. Anderson. 2000. *Structural mechanics of buried pipes*. Boca Raton, FL: CRC Press.

Wylie, E. B., and V. L. Streeter. 1993. *Fluid transients in systems*. Englewood Cliffs, NJ: Prentice Hall.

Index

R

S